Municipal Waste Management: Policies and Strategies

Municipal Waste Management: Policies and Strategies

Edited by

Sanjeev Kumar
Lovely Professional University, India

Mohammad Badruddoza Talukder
International University of Business Agriculture and Technology, Bangladesh

A.K. Haghi
University of Coimbra, Coimbra, Portugal

CABI

CABI is a trading name of CAB International

CABI	CABI
Nosworthy Way	200 Portland Street
Wallingford	Boston
Oxfordshire OX10 8DE	MA 02114
UK	USA
Tel: +44 (0)1491 832111	T: +1 (617)682-9015
E-mail: info@cabi.org	E-mail: cabi-nao@cabi.org
Website: www.cabi.org	

The views expressed in this publication are those of the author(s) and do not necessarily represent those of, and should not be attributed to, CAB International (CABI). Any images, figures and tables not otherwise attributed are the author(s)' own. References to internet websites (URLs) were accurate at the time of writing.

CAB International and, where different, the copyright owner shall not be liable for technical or other errors or omissions contained herein. The information is supplied without obligation and on the understanding that any person who acts upon it, or otherwise changes their position in reliance thereon, does so entirely at their own risk. Information supplied is neither intended nor implied to be a substitute for professional advice. The reader/user accepts all risks and responsibility for losses, damages, costs and other consequences resulting directly or indirectly from using this information.

CABI's Terms and Conditions, including its full disclaimer, may be found at https://www.cabidigital-library.org/terms-and-conditions.

A catalogue record for this book is available from the British Library, London, UK.

ISBN-13: 9781836990642 (hardback)
        9781836990659 (ePDF)
        9781836990666 (ePub)

DOI: 10.1079/9781836990666.0000

Commissioning Editor: Ward Cooper
Editorial Assistant: Theresa Regueira
Production Editor: Rosie Hayden

Typeset by Exeter Premedia Services Pvt Ltd, Chennai, India
Printed in the USA

Contents

_____

Contributors

———————————

Rizal Akbar Aldyan   National Research and Innovation Agency (BRIN), Indonesia.

Nageswara Rao Ambati   Gujarat National Law University (GNLU), India.

Widodo Aribowo   Sebelas Maret University, Indonesia.

Chirra Baburao   Gujarat National Law University (GNLU), India.

Surabhi Bharti   Guru Kashi University, India.

Ayan Chatterjee   Medhavi Skills University, India.

Pyali Chatterjee   Institute of Chartered Financial Analysts of India (ICFAI) University, India.

Mohammad Abu Daud   State University of Bangladesh, Bangladesh.

Sujayalakshmi Devarayasamudram   North Carolina Central University, USA.

Pranjali Gaur   Rajiv Gandhi National University of Law, India.

Chanel Tri Handoko   Sebelas Maret University, Indonesia.

Agung Hidayat   Sebelas Maret University, Indonesia.

Md. Shafiqul Islam   International University of Business Agriculture and Technology, Bangladesh.

Siti Khoiriyah   Sebelas Maret University, Indonesia.

Rinkesh Kumar   Guru Kashi University, India.

Sanjeev Kumar   Lovely Professional University, India.

Alka Maheshwari   Amity University, India.

Geetha Manoharan   SR University, India.

Prerna Mehta   GD Rungta College of Science and Technology, India.

Prakash Chandra Pandey   SRM University, India.

Kiran A. Parida   Gujarat National Law University (GNLU), India.

Sarita Peddi   SR University, India.

Mr Vipan Rai   Lovely Professional University, India.

S.G. Rama Rao   GITAM (Deemed to be University), India.

Ashi Ramavat   SRM University Delhi-NCR, India.

Kolli Srinivasa Rao   GITAM (Deemed to be University), India.

Rohit Saroha   SGT University, India.

Hari Harjanto Setiawan   National Research and Innovation Agency (BRIN), Indonesia.

Om Sharma   Amity University, India.

Amrik Singh   Lovely Professional University, India.

Mohammad Badruddoza Talukder   International University of Business Agriculture and Technology, Bangladesh.

Ivneet Kaur Walia   Rajiv Gandhi National University of Law, India.

Preface

Municipal solid waste management is an environmental issue that has received considerable attention from researchers and environmental managers for decades. This volume was conducted to present the challenges in implementing the policy on solid waste management along with the strategies from the national and local levels.

This book analyzes existing policies related to solid waste management and proposes policy recommendations for effective solid waste management. It discusses implementation strategies and policy adoption, emphasizing the role of policy in shaping and improving solid waste management frameworks.

The journey begins with an introduction to sustainable waste management practices, setting the stage for an in-depth exploration of the strategic waste management landscape. The subsequent chapters cover fundamental aspects, such as current disposal methods and the development of sustainable solutions along with the technological interventions and eco-innovative solutions, laying the groundwork for effective strategic waste management.

This book stands as a significant contribution to the discourse on environmental policy and innovative tools for effective solid and hazardous waste management and our chapters cover a broad spectrum of topics, such as the basics of solid waste generation and new insights into monitoring and adjusting waste strategies.

The contributors to this volume bring a wealth of knowledge and experience, offering original case studies, theoretical models, and practical applications. We are particularly excited to highlight the successful local experiences shared in this book, which offer actionable insights and proven strategies that can be adapted to various areas.

We discuss emerging trends, potential innovations, and the ongoing need for effective policies and regulatory frameworks. By promoting community awareness, leveraging technology, and fostering partnerships between stakeholders, we hope to inspire meaningful progress in waste management practices.

We study the sustainable management of household solid waste (HSW) that explores the cognitive processes behind impactful choices in a strategic waste disposal context. We also look into real-world applications of the latest technologies to control pollution in air, water, and soil by emphasizing the role of creativity and adaptability in achieving sustainable success.

The book is a valuable resource for students, practitioners, and anyone seeking a comprehensive understanding of strategic waste management. It combines theoretical insights with practical applications, making it an indispensable guide for navigating strategic waste disposal's complex and dynamic landscape in the sustainable world.

This volume brings to the fore the fundamentals and practical relevance of a framework for environmental safety and the future of sustainability in the ecosystem. It provides emerging trends and developments in waste management to control contamination in the environment.

This research-oriented book discusses the potential for technological innovations and presents forecasts and predictions for future directions for human health by exploring avenues for advancement in waste management practices. It also provides broad case studies to show the interconnection of environmental pollution with human health and the impact of sustainable development of a healthy environment on proper life standards.

1 Role of Swachhata Entrepreneurs in Municipal Waste Management: Challenges, Opportunities, and Strategic Directions for Sustainability

Chirra Baburao[1]*, Nageswara Rao Ambati[2], Kiran A. Parida[1] and Mohammad Badruddoza Talukder[3]

[1]Gujarat National Law University (GNLU), India; [2]Department of Social Work, Gujarat National Law University, India; [3]International University of Business Agriculture and Technology, Bangladesh

Abstract

This study looks at the role of Swachhata Entrepreneurs in municipal waste management in urban India. In order to respond to growing amounts of waste caused by urbanization, we need innovative solutions. Swachhata Entrepreneurs maximize this potential, turning municipal solid waste into a range of valuable resources through recycling, composting, and energy recovery, in line with the Swachh Bharat Abhiyan campaign and the principles of the circular economy. The research highlights the major challenges encountered by entrepreneurs regarding financial capacity, development infrastructure, public awareness, and gender constraints. But potential markets for recyclables are starting to open up and the public is becoming more aware of sustainability issues, making for a bright future. The study highlights policy actions for supporting Swachhata Entrepreneurs, which includes facilitating access to finance, investing in infrastructure for waste management, awareness generation, promoting gender equality, adopting technological advances and resilient business models, and strengthening policy interventions. However, by addressing these issues and overcoming the obstacle while making use of their opportunities, Swachhata Entrepreneurs can make a positive impact on sustainable waste management and economic growth in urban India.

1.1 Introduction

With urbanization, especially in developing countries, municipal solid waste management is becoming an increasingly challenging problem in India which needs innovative solutions to tackle this looming problem (Kumar *et al.*, 2017). In light of the Swachh Bharat Abhiyan campaign and principles of circular economy,

Swachhata Entrepreneurs can play a vital role when it comes to treating municipal solid waste (MSW) and converting it into commercially competitive resource products (Venkata Mohan *et al.*, 2017; Ferronato *et al.*, 2019; Aggarwal *et al.*, 2020; Fiksel *et al.*, 2020). These entrepreneurs solve urban waste problems and contribute to environmental sustainability and economic development by using practices such

*Corresponding author: bchirra@gnlu.ac.in

© CAB International 2025. *Municipal Waste Management: Policies and Strategies*
(eds S. Kumar *et al.*)
DOI: 10.1079/9781836990666.0001

as recycling, composting, and energy recovery (Binder and Torres, 2009; Chen *et al.*, 2016). They are indispensable in connecting the publicly known data points with adequate actions to attain significant reductions in sustainability-related effects, thus providing the underlying structure and solutions required to implement an efficient waste (Kaza *et al.*, 2016; Fang *et al.*, 2023; Etim, 2024). However, in developing economies, Swachhata Entrepreneurs face multiple challenges including limited availability of finance and resources, lack of infrastructure facilities and prevalence of negative perception about waste-related work among the general population (Lohri *et al.*, 2017; Godfrey *et al.*, 2019). Despite such challenges the circular economy concepts offer vast opportunities associated with the emerging market of recyclable commodities and an increase in the public awareness of waste management superiority (Conlon *et al.*, 2019; Fiksel *et al.*, 2020; Sharma *et al.*, 2021). The Swachhata Entrepreneurs should flourish in the long run in a multi-stakeholder approach. Policymakers and local urban authorities need to undertake specific policy measures, capacity-building schemes, and economic incentives for these entrepreneurs (Creech *et al.*, 2014). It will also aid in creating strong synergy among the public, private, and civic sector building the Swachhata Entrepreneurs' ecosystem and complementing them at providing resources, technology, and expertise to meet their requirements and enhance their functioning and impact (Veleva, 2020). Therefore, this chapter aims to conduct a systematic review of the relevant literature to gain insights into the opportunities and challenges faced by Swachhata Entrepreneurs, with a particular focus on municipal waste management and to outline potential strategic pointers for policymakers, urban local bodies, and other stakeholders to promote entrepreneurship and sustainability in this important area.

1.2 Objective of the Study

This study aims to present a systematic literature review to elucidate the prospects and challenges of Swachhata Entrepreneurs in municipal waste management in urban India.

The specific aims of this study are:

1. To understand the obstacles to the adoption of circular economy practices in municipal waste management businesses run by Swachhata Entrepreneurs.
2. To explore the avenues that Swachhata Entrepreneurs have in converting urban waste into marketable resource products.
3. To increase strategic knowledge of relevant policies, from government policymakers, urban local authorities and other urban stakeholders to promote entrepreneurship and sustainable development in municipal waste management.

1.3 Literature Review

Despite existing literature on the subject, there are still glaring gaps in addressing the scope of engagement with urban waste and the potential entrepreneurs who can effectively address these challenges, especially around Swachhata. Other past studies were either concentrated on the principles of the circular economy without offering clear recommendations to overcome systemic challenges, such as lack of economic viability, inadequate infrastructure, and lack of public awareness, or packaged recommendations that did not translate into action. This emphasizes the importance of enabling Swachhata Entrepreneurs through various interventions such as public–private partnership, gender sensitive approaches, and more. It also highlights the importance of aligning policy frameworks with the realities of the business world.

1.3.1 Addressing the challenges of municipal solid waste management in urban India: the role of Swachhata Entrepreneurs

Municipal Solid Waste Management (MSWM) in urban India today runs up against a multitude of challenges so daunting that they are hard to ignore (Kumar *et al.*, 2017). The context is undeniably complex, with issues ranging from financial barriers and institutional inadequacy to poorly chosen technology and a general public indifference toward segregating and disposing waste properly (Vij, 2012). Though the Swachh Bharat Abhiyan (Clean India Mission) has led to increased awareness and initiatives in this area, there are questions regarding Swachhata Entrepreneurs that require a more nuanced understanding (Charles, 2019; Shukla et al., 2021). By adapting the principles of circular economy Swachhata Entrepreneurs have identified 12 possible issues to encourage change (Goyal et al., 2016). They collaborate with startups to develop recycling and upcycling programs, from a mobile app that encourages the composting of organic food waste to a pilot program that transforms discarded products into high-quality materials. But these entrepreneurs also face some tough obstacles that might limit their overall impact. These challenges should be recognized and addressed by investigating true potential, suitability, and limitations of AI technologies (Veleva, 2020).

1.3.2 Financial constraints and limited access to formal financial systems

One of the major challenges faced by Swachhata Entrepreneurs is access to traditional financial systems (Chen and Divanbeigi, 2019). Securing financing for projects in waste management tends to be an uphill battle. Traditional finance centers are biased toward financing waste management projects due to its perception of risk and uncertainty (Yadav et al., 2010). It is therefore a struggle to get the capital required to grow and to introduce innovative solutions to solve identified problems (Estrin et al., 2018).

1.3.3 Insufficient infrastructure and public awareness

The third major issue is the poor infrastructure for waste management in urban India (Kumar et al., 2017; Aggarwal et al., 2020). It also means that sorting and recycling as well as disposal facilities and systems are still underdeveloped. With little incentive and direction from the local government, Swachhata Entrepreneurs are often left suffering in pursuit of using their human resources effectively in equipment. This inhibits the waste management systems' efficiency and sustainability (Lohri et al., 2017), as these infrastructures are lagging. Moreover, the public still are not aware that they need to sort their waste. Despite all the campaigns geared toward promoting the recycling and the appropriate disposal of waste, individuals who reside in metropolitan areas frequently place their regenerative waste in the same bin as their non-biodegradable waste (Nepal et al., 2022). This lack of public awareness presents a massive issue for entrepreneurs when it comes to collecting, sorting, and recycling waste. Without mass public participation, the platforms generated by Swachhata Entrepreneurs have very little chance of putting in practical or sustainable work (Sabeet, 2017).

1.3.4 The role of government programs and public–private partnerships

Although preceded by multiple government programs, new technological opportunities, and private–public partnerships available to Swachhata Entrepreneurs, such initiatives alone may not be effective in overcoming the existing challenges in the field to a satisfactory level (Goel and Rishi, 2011). Government support may not be stable, and there can be bureaucratic reasons for blocking progress. Furthermore, implementation of new technologies in waste management typically necessitates substantial investment and expertise which might not be accessible to all entrepreneurs (Lohri et al., 2017). Although they can help to make up the shortfalls between government schemes and private sector initiatives (Olukanni and Nwafor, 2019). Nongovernmental organizations (NGOs) can help build a bridge between the needs of the local authorities and private enterprises to plan, implement, and sustain innovative and environmentally responsible solutions to waste management (Godfrey et al., 2019). However, these types of partnerships only flourish under good coordination and a shared commitment, which is not always easy to find.

1.3.5 Market opportunities and sustainability

As demand for recyclable materials increases and public awareness of sustainability grows, new possibilities are being created for Swachhata Entrepreneurs (Veleva, 2020). The growing demand for recycled materials means that entrepreneurs in waste recycling and upcycling can identify potential income streams. Additionally, the growing consumer preference for sustainable products and services is one of the main reasons for convincing companies to become sustainable (Liu *et al.*, 2020). However, these market opportunities, by themselves, are not enough to guarantee Swachhata Entrepreneurs' (Veleva, 2020) sustained growth and success in the future.

Waste management in general: the waste management sector is undoubtedly the high stakes playing arena for entrepreneurship and requires a deep understanding of the industry (Kumar *et al.*, 2017). Furthermore, the economic success and sustainability of waste management initiatives can be influenced by market dynamics and economic volatility. For sustainable growth, Swachhata Entrepreneurs will need to build business models that are adaptive to changing customer consumer needs and resilient to market fluctuations (Bagire *et al.*, 2020).

1.3.6 Gender equality and societal barriers

It is well documented in the literature that gender equality can be critical for enhancing entrepreneurship and innovation for waste management (Outsios and Farooqi, 2017; Charles, 2019; Butkouskaya *et al.*, 2020; Seager *et al.*, 2020; Barrachina *et al.*, 2021; Srinivas, 2022). In this regard, women-led Swachhata Enterprises have worked to catalyze local solutions for the waste they produce, the empowerment of their communities and integrated sustainable development (Chaturvedi, 2003). However, such ventures frequently face broader societal and structural constraints on their development and success (Singh *et al.*, 2020).

Food and economy: in developing countries such as India, some of the biggest hindrances faced by women entrepreneurs are gender bias, lack of access to education and training, and limited mobility (Kothari, 2017; Mahadi *et al.*, 2017; Sowmya and Mishra, 2019; Kumar and Singh, 2021). In addition, challenges may also further permeate through lack of access to funding, networks, and markets. These constraints need to be resolved to unlock the possibilities for these women-led Swachhata Enterprises, which would necessitate a gender-enabling ecosystem.

1.3.7 A nuanced and comprehensive approach

It requires a more nuanced and holistic approach to the promotion of Swachhata Entrepreneurs, so that they could sustain and succeed in their growth (Sahoo, 2020). Efforts should concentrate on resolving the structural systematic problems and the interrelated web of social, economic, and political environment affecting the waste management landscape in the urban India (Ganesan, 2019; Mokale, 2019). Such an enabling ecosystem is necessary to develop for entrepreneurial innovation and sustainable waste management practices to thrive through the involvement of policymakers, stakeholders, and communities (Fiksel *et al.*, 2020).

1.3.8 The path forward

In this sense the circular economy faces a very rich and open future, however, the Swachhata Entrepreneurs' future is full of challenges and uncertainties (Conlon *et al.*, 2019). Kumar (2013) also states that entrepreneurs must approach the challenge in a holistic way, integrating such disparate issues as financial sustainability with infrastructure development, public awareness, gender parity, and responsiveness to the market. Identifying such a multitude of problems, Swachhata Entrepreneurs constitute an important pillar of the transforming urban waste management ecosystem for a cleaner and sustainable India.

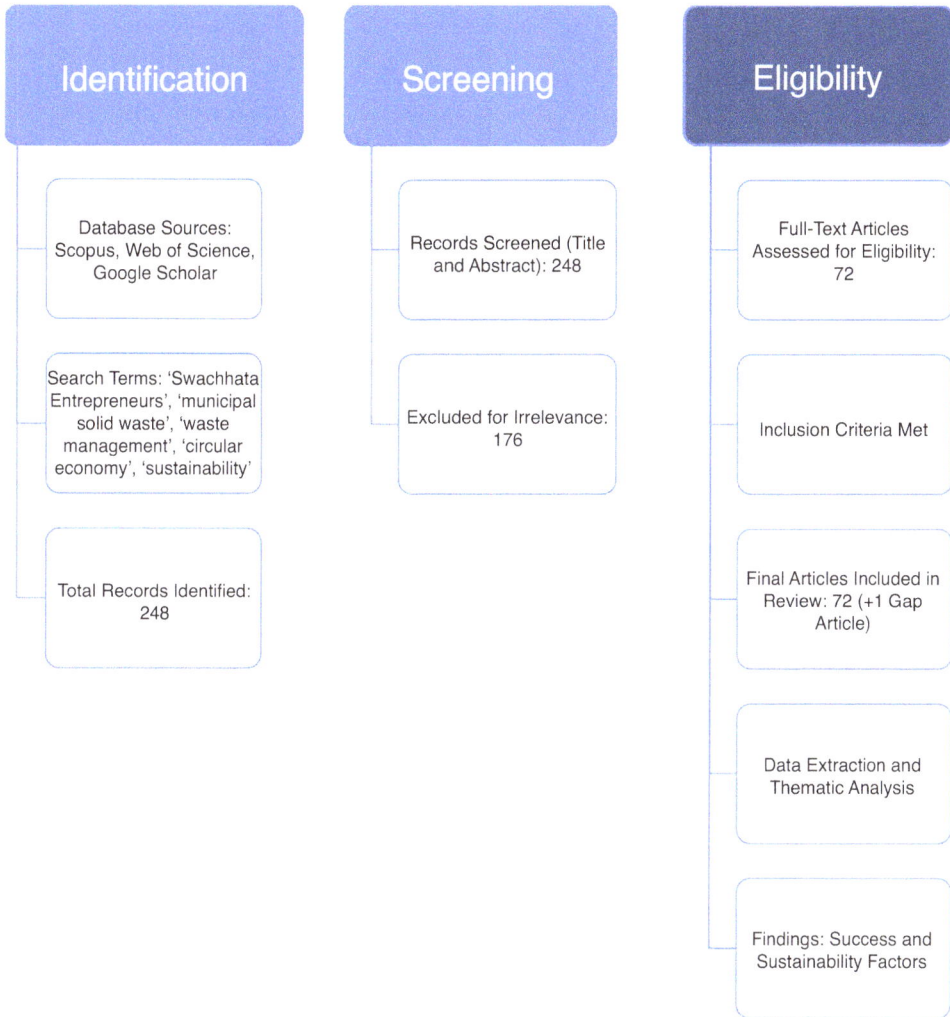

```
┌─────────────────┐   ┌─────────────────┐   ┌─────────────────┐
│  Identification │   │   Screening     │   │   Eligibility   │
└─────────────────┘   └─────────────────┘   └─────────────────┘

  Database Sources:       Records Screened (Title    Full-Text Articles
  Scopus, Web of Science, and Abstract): 248         Assessed for Eligibility:
  Google Scholar                                      72

  Search Terms: 'Swachhata
  Entrepreneurs', 'municipal  Excluded for Irrelevance:  Inclusion Criteria Met
  solid waste', 'waste         176
  management', 'circular
  economy', 'sustainability'

  Total Records Identified:                           Final Articles Included in
  248                                                 Review: 72 (+1 Gap
                                                       Article)

                                                       Data Extraction and
                                                       Thematic Analysis

                                                       Findings: Success and
                                                       Sustainability Factors
```

Fig. 1.1. Systematic literature review approach. Author's own.

1.4 Methodology

The methodology summarized in Fig. 1.1 employs a systematic literature review approach to examine issues, opportunities, and strategic directions surrounding Swachhata Entrepreneurs and municipal waste disposal in urban Indian areas (Shukla *et al.*, 2021). We conducted a systematic literature review (SLR) of peer-reviewed journal articles, conference proceedings, and gray literature from leading databases such as Scopus and Web of Science (Mani and Singh, 2016). The search query was constructed to aggregate context-related literature about the government body, Swachhata Entrepreneurs, cognizant of the municipal waste management challenges that they were addressing, as well as terms such as Swachhata Entrepreneurs, municipal solid waste, waste management, circular economy, and sustainability (Gittins, 2020).

The first search yielded 248 articles, which were screened by title and abstract when relevant. This resulted in 72 articles being reviewed in full (Sulemana *et al.*, 2018). After going through the relevant literature, the final selection criteria included the fact that articles: (i) should describe the role that Swachhata Entrepreneurs or such waste management enterprises have held in urban India; (ii) should cover their challenges, opportunities, and strategic ways forward; (iii) should be written in English and published between 2020 and 2024 (Chaturvedi, 2003).

The researchers identified a research gap and subsequently used an article from Google Scholar to understand the gap-defining challenges, opportunities, and strategic recommendations for Swachhata Entrepreneurs (Kraus *et al.*, 2018). Any variation in extracted data was discussed and agreed on. Data extraction used thematic analysis to synthesize common themes and patterns shown across the selected articles. The themes were additionally condensed and bucketed to provide insight on the factors driving the success and sustainability of Swachhata Entrepreneurs within the municipal waste management domain (Ghosh, 2016).

1.5   Findings and Results

The study yielded several significant insights into the function of Swachhata Entrepreneurs in urban India's municipality waste management. These results demonstrate the opportunities and challenges that these entrepreneurs encounter and also the strategic avenues that might sustain their long-term viability.

1.5.1   Opportunities for Swachhata Entrepreneurs

Rising interest among principles of circular economy and sustainability provides fresh channels of opportunity for Swachhata Entrepreneurs (Goyal *et al.*, 2016). In the field of waste recycling and upcycling, there is a, even more, lucrative income opportunity, due to the growing demand for recycled materials. These are processes in which waste is made into raw material that we can use in an economically viable way to reduce the environmental burden (Ranta and Saari, 2019).

Achieving circular economy: Moreover, the preferences of consumers are slowly changing toward sustainable products and services. In light of increasing awareness of environmental issues, increasing numbers of people are looking to green alternatives (Serrano-Arcos *et al.*, 2021). This has led to a shift in consumer preferences contributing in turn to adopting sustainable practices, encouraging Swachhata Entrepreneurs to offer new products and services (Geeta, 2023).

Yet, these market opportunities are unfulfilled and will not necessarily translate to Swachhata Entrepreneurs' long-term growth and success. The waste management sector is also a competitive industry and entrepreneurs are always compelled to identify innovative strategies against dynamic markets (Bagire *et al.*, 2020). As such, entrepreneurs need to keep up with developments in technology and the market so as not to fall behind competitors and to ensure that they remain useful in their practice (Shahidullah and Haque, 2014). Waste management businesses can be affected by market forces and economic uncertainties. For sustainable growth, Swachhata Entrepreneurs must develop robust business models that are evergreen and can handle market fluctuations (Salana, 2019). Then they will be resilient and able to navigate challenges and take advantage of budding opportunities.

1.5.2   Challenges and barriers

The thorough literature review revealed some of the chronic challenges and hindrances facing Swachhata Entrepreneurs operating in the municipal solid waste management sector.

Swachhata Entrepreneurs face significant challenges when it comes to funding their businesses. The banks and financial institutions may be very reluctant to lend money to waste management start-ups because of the underlying risks and uncertainties (Charles, 2019). It has also made it more difficult for entrepreneurs to quickly access capital needed for infrastructure, equipment, and other operating expenses. Funding is the key to expanding operations and adopting creative solutions (Buteau, 2021).

Several cities in India do not have the infrastructure for efficient collection of municipal solid waste (Vij, 2012), such as collection systems, processing facilities, and disposal sites. Lack of infrastructure creates logistical or operational hurdles for Swachhata Entrepreneurs to collect, segregate, and process waste in a timely manner. These obstacles are faced by the bulk of these entrepreneurial entities, which points to the need for a strong waste management infrastructure to be built and supported (Ghahramani et al., 2021).

Waste segregation and recycling initiatives are typically characterized by poor public awareness and participation. Many people in urban areas fail to separate biodegradable and non-biodegradable waste despite campaigns for practicing responsible waste disposal practice (Alam et al., 2007). This lack of awareness serves as one of the biggest hurdles for these entrepreneurs as it makes the collection, segregation, and recycling of waste extremely complicated. Swachhata Entrepreneurs' success depends on changing the population's behavior and raising public awareness (Charles, 2019).

Previous studies show that gender equality in waste has positive implications for entrepreneurship and innovation. Women-led Swachhata Enterprises have shown potential promise in tackling local waste issues, improving the livelihood of marginalized groups, and advancing sustainable development (Chaturvedi, 2003). These enterprises tend to face systemic societal and cultural barriers that exclude them from key resources, networks, and even decision-making power. When negotiating gender-based obstacles and advancing gender parity, these endeavors should be targeted from the perspective of realizing the full capability of women-driven Swachhata Enterprises (Murthy, 2010).

1.5.3 Strategic directions for sustainability

In order to overcome the challenges and capitalize on the opportunities identified, the study proposes the following strategic directions for ensuring the long-term viability of the Swachhata Entrepreneurs engaged in municipal waste management (Bagire et al., 2020).

Access to financing is crucial for the growth and success of Swachhata Entrepreneurs. From financial agencies' standpoint, this should be focused on targeted funding and grants for waste management startups (Salana, 2019). Also, by raising awareness for available financing options, along with providing financial literacy training, entrepreneurs will be better equipped to access the capital they require for their businesses. Such integration will help create a conducive environment to the Swachhata Entrepreneurs and therefore it will require an investment on this front to develop a user-friendly waste management infrastructure.

Governments and local authorities: focus on efficient collection systems, processing facilities, and disposal sites so that public–private partnerships are a key player for infrastructure development resource-sharing and capacity building (Joshi and Ahmed, 2016; Kumar et al., 2017).

Swachhata Entrepreneurs would greatly benefit from public awareness campaigns on waste segregation and recycling. Governments NGOs and community organizations' health education should be focused on such strategies (Afroz and Masud, 2010). These initiatives are designed to raise the awareness of residents on the importance of disposing waste responsibly and should act as a call to action. In recent years, there has been a concerted effort in the waste management industry to help economically empower women entrepreneurs and address gender equality issues. Policymakers and other stakeholders can bring gender-specific barriers down by enabling women-led Swachhata Enterprises. This will include access to education and training, networking opportunities, as well as representation in the decision-making framework (Nhamo and Mukonza, 2020).

As a result, data-driven models can be developed which can provide invaluable insights and recommendations to empower entrepreneurs to optimize their waste management operations.

Smart breaking solutions: Innovative technologies such as waste-to-energy, advanced recycling, and new formats for waste management and tracking moves to digital platforms become increasingly relevant for Swachhata Entrepreneurs, these innovative technologies will also play a substantial role in the Swachhata economy. Specific funding and partnership provision would incentivize the implementation of these types of technologies, and national and research organizations need to support research and development of such technologies (Borchard *et al.*, 2021). It is critical for Swachhata Entrepreneurs to create sustainable business models that can adapt to changing market dynamics in the long term. Adaptability skill entrepreneurs also need to continuously monitor market trends, consumer preferences, and regulatory shifts to remain competitive (Veleva, 2020; Gyimah and Adeola, 2021). This diversification strategy allows them to mitigate the risks associated with market fluctuations and economic uncertainties.

For Swachhata Entrepreneurs to flourish, they also need appropriate policy support and advocacy in place. Policymakers should create and implement comprehensive waste generation policies that guarantee entrepreneurship, creativity, and sustainability. This may help advocate policy changes and concessions that accommodate the specific needs and challenges of Swachhata Entrepreneurs from different sectors, depending on their geographical location, as well as based on their scale of operations. These entrepreneurs have immense potential, but they also present significant challenges that require targeted interventions.

Participative solutions to tackle challenges of Swachhata Entrepreneurs: financial barriers, infrastructure challenges, limited awareness, harder gender barriers, collaboratives and market barriers can be resolved by using participative solutions which will ensure that stakeholders can build an enabling ecosystem which can lead to sustained growth and success of Swachhata Entrepreneurs.

1.6 Discussion

This study can be orienting toward the in-depth focus on Swacchata Entrepreneurs and their vast impact on tackling the ever-mounting municipal waste problem in urban India. Inspired by the Swachh Bharat Abhiyan and principles of circular economy, some of these entrepreneurs have been able to generate commercially sustainable resources from waste (Fiksel *et al.*, 2020). The report also highlights that Swachhata Entrepreneurs are a reality and

an unparalleled opportunity in the domain of waste management. These entrepreneurs help to save the planet by organizing recyclable and organic waste into innovative products, they are also boosting economic growth by adopting sustainable practices such as recycling, composting, and energy recovery (Kumar and Sil, 2015; Etim, 2024). Waste reuse could alleviate pressure on municipal waste systems, create jobs, and enable a more resource efficient and circular economy (Godfrey *et al.*, 2019). But the study also finds significant hurdles that Swachhata Entrepreneurs face in their work. A few of these challenges are the need of keeping financing available, not having enough waste management plants, and low general public awareness with regards to the segregation of waste. These challenges are what typically hinder the entrepreneurs from growing their operations and scaling sustainably long term.

Based on the above challenges and opportunities for the sustainability of Swachhata Entrepreneurs due to its immense potential and significance, the study recommends the following.

1. Facilitating access to financing: by creating specialized funding opportunities and incentives for entrepreneurs, alongside educational initiatives to enhance their financial capacity, these financial institutions can help entrepreneurs secure the funds they need to expand their businesses.
2. Investing in waste management infrastructure: governments and local authority should prioritize an efficient collection system and processing facility and disposal site. This is where public–private partnerships can be an essential part of the solution.
3. Incralso work together to raise awareness of the importance of waste segregation and recycling through educational campaigns and outreach programs.
4. Promoting gender equality: by ensuring the supply side considering women by addressing the gender-based barriers and empowering women entrepreneurs for the waste management sector to unlock their full potential by adoption of Swachhata Enterprises contributing to sustainable development.

5. Embracing technological advancements: utilizing new technologies, including but not limited to waste-to-energy conversion as well as advanced recycling methods, can help improve the efficiency and effectiveness of waste management activities.

6. Developing resilient business models: waste is unavoidable, but building up a concrete framework for such policies can ensure that growth leads to a more sustainable and waste-free future for the economy.

7. Streasing public awareness: governments, NGOs, and community organizations can engthening policy support and advocacy: as such, policymakers and stakeholders should partner to introduce comprehensive waste management policies that encourage entrepreneurship, innovation, and sustainability.

The study identifies (as shown in Fig. 1.2) these strategic directions which stakeholders may adopt to develop an enabling ecosystem for building Swachhata Entrepreneurs that thrive. The entrepreneurs, the India Urban WASH (water, sanitation, and hygiene) community, can the support cleaner environmental conditions, economic opportunity, and the realization of sustainable development goals for urban India.

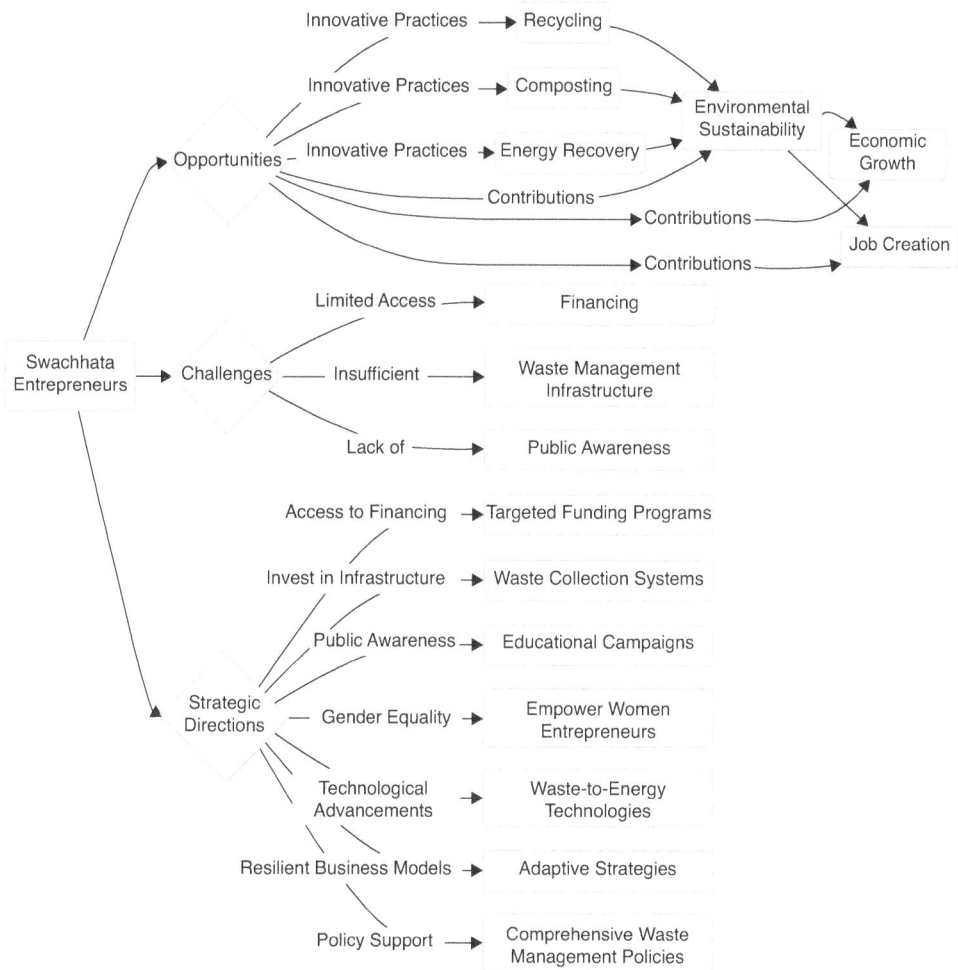

Fig. 1.2. Framework for Swachhata Entrepreneurs in Waste Management. Author's own.

1.7   Conclusion

Inspired by the Swachh Bharat Abhiyan movement and following the principles of a circular economy, Swachhata Entrepreneurs can be a game changer in addressing the ever-increasing problem of municipal solid waste management in urban India. This study provides significant insights into these entrepreneurs' role in the management of municipal solid wastes and can accordingly be of interest to policymakers, urban local authorities, and risk bearers. The insights gained from such case studies may provide valuable guidance for researchers, practitioners, and policymakers aiming to understand the unique dynamics and hurdles faced by these micro and small enterprises, and ultimately support their growth and resilience.

1.8   Future Research

This study provides a basis for understanding the role of Swachhata Entrepreneurs in the municipal solid waste management in the urban regions of India. But behind these lessons lies a logic that future research can expand: case studies that unwrap the particular box of challenges, strategies, and best practices that have worked for successful Swachhata Enterprises; comparative studies that illuminate the shades of Swachhata entrepreneurship across cultures and economies; impact evaluations of government policies and programs that can guide policymakers to adopt the most effective approaches; identification of next generation business models that leverage digital technologies or combine the informal sector; and digging into gender barriers with in-depth research on women waste entrepreneurs, how policy can actively promote their participation and success. This will help guide policymakers, urban local bodies, and others to in turn create a virtuous cycle for Swachhata Entrepreneurs, so that these innovative enterprises are enabled to be sustainable and for their services to be scaled up.

References

Afroz, R. and Masud, M.M. (2010) Using a contingent valuation approach for improved solid waste management facility: Evidence from Kuala Lumpur, Malaysia. *Waste Management* 31(4), 800–808. DOI: 10.1016/j.wasman.2010.10.028.

Aggarwal, L., Aggarwal, V., Aggarwal, R. and Aggarwal, V.K. (2020) On an attempt to explore the possible challenges faced during the solid waste management using ISM methodology. *International Journal of Computer Applications* 177(43), 55–60. DOI: 10.5120/ijca2020919948.

Alam, R., Chowdhury, M.A.I., Hasan, G.M.J., Karanjit, B. and Shrestha, L.R. (2007) Generation, storage, collection and transportation of municipal solid waste. *Waste Management* 28, 1088–1097. DOI: 10.1016/j.wasman.2006.12.024.

Bagire, V., Wafler, M., Rieck, C.E., Asiimwe, J., Abaho, E. *et al.* (2020) Waste as business: Emerging Ugandan micro- and small-sized businesses in resource recovery and safe reuse. *Journal of Environmental Management* 279, 111802. DOI: 10.1016/j.jenvman.2020.111802.

Barrachina, M., Centeno, M.C.G. and Patier, C.C. (2021) Women sustainable entrepreneurship: Review and research agenda. *Sustainability* 13(21), 12047. DOI: 10.3390/su132112047.

Binder, J.J. and Torres, S.A. (2009) Implementation of the Taunton, Massachusetts regional solid waste management facility. In: *Proceedings of the 17th Annual North American Waste-to-Energy Conference*, Chantilly, VA, May 18–20, pp. 101–106. DOI: 10.1115/nawtec17-2310.

Borchard, R., Zeiss, R. and Recker, J. (2021) Digitalization of waste management: Insights from German private and public waste management firms. *Waste Management and Research The Journal for a Sustainable Circular Economy* 40(6), 775–792. DOI: 10.1177/0734242x211029173.

Buteau, S. (2021) Roadmap for digital technology to foster India's MSME ecosystem – opportunities and challenges. *CSI Transactions on ICT* 9(4), 233–244. DOI: 10.1007/s40012-021-00345-4.

Butkouskaya, V., Romagosa, F. and Noguera, M. (2020) Obstacles to sustainable entrepreneurship amongst tourism students: A gender comparison. *Sustainability* 12(5), 1812. DOI: 10.3390/su12051812.

Charles, G. (2019) Sustainability of social enterprises involved in waste collection and recycling activities: Lessons from Tanzania. *Journal of Social Entrepreneurship* 12(2), 219–237. DOI: 10.1080/19420676.2019.1686712.

Chaturvedi, B. (2003) Waste-handlers and recycling in urban India: Policy, perception and the law. *Social Change* 33, 41–50. DOI: 10.1177/004908570303300304.

Chen, P., Xie, Q., Addy, M., Zhou, W., Liu, Y. *et al.* (2016) Utilization of municipal solid and liquid wastes for bioenergy and bioproducts production [Review of utilization of municipal solid and liquid wastes for bioenergy and bioproducts production]. *Bioresource Technology* 215, 163–172. DOI: 10.1016/j.biortech.2016.02.094.

Chen, R. and Divanbeigi, R. (2019) Can regulation promote financial inclusion? policy research working paper 8711, world bank. DOI: 10.1596/1813-9450-8711.

Conlon, K., Jayasinghe, R. and Dasanayake, R. (2019) Circular economy: Waste-to-wealth, jobs creation, and innovation in the global south. *World Review of Science Technology and Sustainable Development* 15(2), 145–159. DOI: 10.1504/wrstsd.2019.099377.

Creech, H., Paas, L., Gabriel, G.H., Voora, V., Hybsier, C. *et al.* (2014) Small-scale social-environmental enterprises in the green economy: Supporting grassroots innovation. *Development in Practice* 24(3), 366–378. DOI: 10.1080/09614524.2014.899561.

Estrin, S., Gozman, D. and Khavul, S. (2018) The evolution and adoption of equity crowdfunding: Entrepreneur and investor entry into a new market. *Small Business Economics* 51(2), 425–439. DOI: 10.1007/s11187-018-0009-5.

Etim, E. (2024) Leveraging public awareness and behavioural change for entrepreneurial waste management. *Heliyon* 10(21), e40063. DOI: 10.1016/j.heliyon.2024.e40063.

Fang, B., Yu, J., Chen, Z., Osman, A.I., Farghali, M. *et al.* (2023) Artificial intelligence for waste management in smart cities: A review. *Environmental Chemistry Letters* 21(4), 1959–1989. DOI: 10.1007/s10311-023-01604-3.

Ferronato, N., Ragazzi, M., Portillo, M.A.G., Lizarazu, E.G.G., Viotti, P. *et al.* (2019) How to improve recycling rate in developing big cities: An integrated approach for assessing municipal solid waste collection and treatment scenarios. *Environmental Development* 29, 94–110. DOI: 10.1016/j.envdev.2019.01.002.

Fiksel, J., Sanjay, P. and Raman, K. (2020) Steps toward a resilient circular economy in India. *Clean Technologies and Environmental Policy* 23(1), 203–218. DOI: 10.1007/s10098-020-01982-0.

Ganesan, P. (2019) Waste of a nation: Garbage and growth in India. *International Journal of Environmental Studies* 76(3), 526–528. DOI: 10.1080/00207233.2019.1568760.

Geeta, J. (2023) Green initiatives of Indian startups to achieve sustainability – a step forward. *International Journal of Engineering Technology and Management Sciences* 7(2), 884–888. DOI: 10.46647/ijetms.2023.v07i02.099.

Ghahramani, M., Zhou, M., Mölter, A. and Pilla, F. (2021) IoT-based route recommendation for an intelligent waste management system. *IEEE Internet of Things Journal* 9(14), 11883–11892. DOI: 10.1109/jiot.2021.3132126.

Ghosh, S.K. (2016) Swachhaa Bharat Mission (SBM) – a paradigm shift in waste management and cleanliness in India. *Procedia Environmental Sciences* 35, 15–27. DOI: 10.1016/j.proenv.2016.07.002.

Gittins, T. (2020) Development of an entrepreneurship typology for integration of Roma informal waste collection practices into environmental policy in the CEE region. *Small Enterprise Research* 27(3), 289–305. DOI: 10.1080/13215906.2020.1835706.

Godfrey, L., Ahmed, M.T., Gebremedhin, K.G., Katima, J.H.Y., Oelofse, S. *et al.* (2019) Solid waste management in Africa: Governance failure or development opportunity? In: Edomah, N. (ed.) *Regional Development in Africa. IntechOpen*. London, pp. 235–248. DOI: 10.5772/intechopen.86974.

Goel, G. and Rishi, M. (2011) Promoting entrepreneurship to alleviate poverty in India: An overview of government schemes, private-sector programs, and initiatives in the citizens' sector. *Thunderbird International Business Review* 54(1), 45–57. DOI: 10.1002/tie.21437.

Goyal, S., Esposito, M. and Kapoor, A. (2016) Circular economy business models in developing economies: Lessons from India on reduce, recycle, and reuse paradigms. *Thunderbird International Business Review* 60(5), 729–740. DOI: 10.1002/tie.21883.

Gyimah, P. and Adeola, O. (2021) MSMEs sustainable prediction model: A three-sector comparative study. *Journal of the International Council for Small Business* 2(2), 90–100. DOI: 10.1080/26437015.2021.1881933.

Joshi, R. and Ahmed, S. (2016) Status and challenges of municipal solid waste management in India: A review. *Sustainable Environment* 2(1), 1139434. DOI: 10.1080/23311843.2016.1139434.

Kaza, S., Yao, L. and Stowell, A. (2016) *Sustainable Financing and Policy Models for Municipal Composting.* World Bank, Washington, DC. DOI: 10.1596/26286.

Kothari, T. (2017) Women entrepreneurs' path to building venture success: Lessons from India. *South Asian Journal of Business Studies* 6(2), 118–141. DOI: 10.1108/sajbs-03-2016-0021.

Kraus, S., Burtscher, J., Vallaster, C. and Angerer, M. (2018) Sustainable entrepreneurship orientation: A reflection on status-quo research on factors facilitating responsible managerial practices. *Sustainability* 10(2), 444. DOI: 10.3390/su10020444.

Kumar, A. (2013) Women entrepreneurs in a masculine society: Inclusive strategy for sustainable outcomes. *International Journal of Organizational Analysis* 21(3), 373–384. DOI: 10.1108/ijoa-01-2013-0636.

Kumar, S. and Singh, N. (2021) Entrepreneurial prospects and challenges for women amidst COVID-19: A case study of Delhi, India. *Fulbright Review of Economics and Policy* 1(2), 205–226. DOI: 10.1108/frep-09-2021-0057.

Kumar, S. and Sil, A. (2015) Challenges and opportunities in SWM in India: A perspective. In: Dev, S. and Yedlah, S. (eds) *Cities and Sustainability.* Springer, New Delhi, India, pp. 193–210. DOI: 10.1007/978-81-322-2310-8_10.

Kumar, S., Smith, S.R., Fowler, G., Velis, C.A., Kumar, S.J. *et al.* (2017) Challenges and opportunities associated with waste management in India. *Royal Society Open Science* 4(3), 160764. DOI: 10.1098/rsos.160764.

Liu, G., Zhang, C., Zhao, M., Guo, W. and Luo, Q. (2020) Comparison of nanomaterials with other unconventional materials used as additives for soil improvement in the context of sustainable development: A review. *Nanomaterials* 11(1), 15. DOI: 10.3390/nano11010015.

Lohri, C.R., Diener, S., Zabaleta, I., Mertenat, A. and Zurbrügg, C. (2017) Treatment technologies for urban solid biowaste to create value products: A review with focus on low- and middle-income settings. *Reviews in Environmental Science and Bio/Technology* 16(1), 81–130. DOI: 10.1007/s11157-017-9422-5.

Mahadi, H.M., Mamun, H. and Saleheen, F. (2017) ICT-based business initiatives for women: An outline of best practices in e-commerce/e-retailing ventures. *Frontiers in Management Research* 1(1), 31–36. DOI: 10.22606/fmr.2017.11005.

Mani, S. and Singh, S. (2016) Sustainable municipal solid waste management in India: A policy agenda. *Procedia Environmental Sciences* 35, 150–157. DOI: 10.1016/j.proenv.2016.07.064.

Mokale, P. (2019) Smart waste management under smart city mission – its implementation and ground realities. *International Journal of Innovative Technology and Exploring Engineering* 8(12), 3095–3103. DOI: 10.35940/ijitee.k1311.1081219.

Murthy, P. (2010) Challenges faced by women entrepreneurs in globalized era. Working Paper, SSRN. DOI: 10.2139/ssrn.1650583.

Nepal, M., Nepal, A.K., Khadayat, M.S., Kumar, R., Shyamsundar, P. *et al.* (2022) Low-cost strategies to improve municipal solid waste management in developing countries: Experimental evidence from Nepal. *Environmental and Resource Economics* 84, 729–752. DOI: 10.1007/s10640-021-00640-3.

Nhamo, G. and Mukonza, C. (2020) Opportunities for women in the green economy and environmental sectors. *Sustainable Development* 28(4), 823–832. DOI: 10.1002/sd.2033.

Olukanni, D.O. and Nwafor, C.O. (2019) Public-private sector involvement in providing efficient solid waste management services in Nigeria. *Recycling* 4(2), 19. DOI: 10.3390/recycling4020019.

Outsios, G. and Farooqi, S.A. (2017) Gender in sustainable entrepreneurship: Evidence from the UK. *Gender in Management* 32(3), 183–202. DOI: 10.1108/gm-12-2015-0111.

Ranta, V. and Saari, U.A. (2019) Circular economy: Enabling the transition towards sustainable consumption and production. In: Leal Filho, W., Azul, A., Brandli, L., Özuyar, P. and Wall, T. (eds) *Responsible Consumption and Production: Encyclopedia of the UN Sustainable Development Goals.* Springer, Cham, Switzerland, pp. 78–89. DOI: 10.1007/978-3-319-71062-4_3-1.

Sabeet (2017) Sustainable solid waste management in India. *International Journal of Scientific Research in Science and Technology* 3(8), 1317–1324. DOI: 10.32628/ijsrst18419.

Sahoo, C. (2020) Women entrepreneurship in India: An insight into problems, prospects and development. *International Journal of Engineering Research and Technology* 9(9), 586–591. DOI: 10.17577/ijertv9is090224.

Salana, S. (2019) Worldwide business models – environmental and financial aspects. In: Kumar, S., Kumar, R. and Pandy, A. (eds) *Current Developments in Biotechnology and Bioengineering*. Elsevier, Amsterdam, pp. 319–345. DOI: 10.1016/b978-0-444-64083-3.00016-6.

Seager, J., Rucevska, I. and Schoolmeester, T. (2020) Gender in the modernisation of waste management: Key lessons from fieldwork in Bhutan, Mongolia, and Nepal. *Gender and Development* 28(3), 551–569. DOI: 10.1080/13552074.2020.1840155.

Serrano-Arcos, M.M., Sánchez-Fernández, R. and Pérez-Mesa, J.C. (2021) Analysis of product-country image from consumer's perspective: The impact of subjective knowledge, perceived risk and media influence. *Sustainability* 13(4), 2194. DOI: 10.3390/su13042194.

Shahidullah, A.K.M. and Haque, C.E. (2014) Environmental orientation of small enterprises: Can microcredit-assisted microenterprises be 'green'? *Sustainability* 6(6), 3232–3251. DOI: 10.3390/su6063232.

Sharma, H.B., Vanapalli, K.R., Samal, B., Cheela, V.R.S., Dubey, B. *et al.* (2021) Circular economy approach in solid waste management system to achieve UN-SDGs: Solutions for post-COVID recovery. *Science of The Total Environment* 800, 149605. DOI: 10.1016/j.scitotenv.2021.149605.

Shukla, P., Sharma, P.K., Pandey, S. and Chintala, V. (2021) Unsegregated municipal solid waste in India – current scenario, challenges and way forward. *Nature Environment and Pollution Technology* 20(2), 851–863. DOI: 10.46488/nept.2021.v20i02.048.

Singh, M., Rathi, R. and Garza-Reyes, J.A. (2020) Analysis and prioritization of Lean Six Sigma enablers with environmental facets using best worst method: A case of Indian MSMEs. *Journal of Cleaner Production* 279, 123592. DOI: 10.1016/j.jclepro.2020.123592.

Sowmya, C.S. and Mishra, J. (2019) Challenges of women entrepreneurs in India and the emerging dimensions: A study. Working paper, SSRN. DOI: 10.2139/ssrn.3555613.

Srinivas, H. (2022) Gender and e-waste: Policy considerations for developing countries. Policy analysis series e-124, Kobe, Japan. Available at: https://gdrc.org/gender/gender-ewaste/index.html (accessed 24 June 2025).

Sulemana, A., Donkor, E.A., Forkuo, E.K. and Oduro-Kwarteng, S. (2018) Optimal routing of solid waste collection trucks: A review of methods. *Journal of Engineering* 1, 4586376. DOI: 10.1155/2018/4586376.

Veleva, V. (2020) The role of entrepreneurs in advancing sustainable lifestyles: Challenges, impacts, and future opportunities. *Journal of Cleaner Production* 283, 124658. DOI: 10.1016/j.jclepro.2020.124658.

Venkata Mohan, S., Chiranjeevi, P., Dahiya, S. and Naresh Kumar, A. (2017) Waste derived bioeconomy in India: A perspective. *New Biotechnology* 40, 60–69. DOI: 10.1016/j.nbt.2017.06.006.

Vij, D. (2012) Urbanization and solid waste management in India: Present practices and future challenges. *Procedia – Social and Behavioral Sciences* 37, 437–447. DOI: 10.1016/j.sbspro.2012.03.309.

Yadav, I.C., Devi, N.L., Singh, S. and Prasad, A.G.D. (2010) Evaluating financial aspects of municipal solid waste management in Mysore City, India. *International Journal of Environmental Technology and Management* 13, 302–310. DOI: 10.1504/ijetm.2010.038009.

2 Solid Waste Management: Environment, Health, and Safety

Surabhi Bharti* and Rinkesh Kumar
Guru Kashi University, India

Abstract

The ways in which waste, whether it is simply household waste or waste from an industry, is disposed of are often not sustainable. Solid waste may be defined as the generation of undesirable substances which is left after they are used once. They cannot be reused directly by society because some of them may be hazardous to human health and welfare. Factors such as rapid urbanization changes in consumption patterns and lack of recycling activities resulted in an increase in waste generation. Improper solid waste management leads to substantial negative environmental impacts and health and safety problems. Solid waste management includes all aspects of waste storage, collection, transportation, sorting, disposal, thermal treatment, composting, and related management required for proper disposal of waste generated at domestic, industrial, commercial, as well as waste generated by institutions, all of which are to be discussed in this chapter. It will also include all the aspects of waste management, its types, quantity, and composition of the waste generated. Solid waste has an adverse effect on public health and it affects the surrounding environment. The actions taken to eradicate problems related to solid waste management with effective solutions that require less investment should be implemented by the policymakers so that the problem can be solved and the impact of these toxic wastes generated can be minimized.

2.1 Introduction

Solid waste is rejected solid or semi-solid material from neighborhood activities, including industrial, agricultural, and commercial sources (Wu et al., 2022). It includes packaging, food scraps, chemicals, sewage, hazardous substances, bulky items like furniture, electronics, and construction debris, divided into organic and inorganic (Arafat et al., 2015) waste which results from human evolution, linked to infrastructure, economy, and lifestyle.

China, India, and the United States lead in municipal solid waste production, with India in third place globally (Nanda and Berruti, 2021). India produces 62 million tons of municipal solid waste (MSW) annually; 43 million tons are collected, but only 12 million tons are treated, while the rest goes to landfills, harming the environment (Mohanty et al., 2022). Waste rises yearly by about 5%, worsening management issues. Globally, 2.01 billion metric tons of MSW are generated yearly with an average of 0.74 kg/day per person (Kaza et al., 2018; Lino et al., 2023). In 2050, global waste could reach 3.4 billion metric tons, with rich nations (16% of the population) generating 34% of waste. Urgent changes are needed to prevent low-income countries from tripling waste and facing increased risks (Tayeh et al., 2020).

Solid waste management is crucial for environmental and public health by preventing

*Corresponding author: drsurabhi.171446@gku.ac.in

© CAB International 2025. *Municipal Waste Management: Policies and Strategies* (eds S. Kumar *et al.*)
DOI: 10.1079/9781836990666.0002

pollution and reducing methane emissions (Abubakar *et al.*, 2022). Traditional landfill and incineration methods harm the environment. Developing countries struggle with waste management, often burning waste, worsening air quality, and affecting nearby communities. Waste burning in landfills releases harmful pollutants, especially in poorer communities which are situated closer to these landfills (Abubakar *et al.*, 2022). Modern practices emphasize sustainability through the 3Rs – reduce, reuse, recycle – and integrated solid waste management. Developing countries face challenges due to limited resources, infrastructure, and governance, impacting public health. A circular economy approach is essential for sustainability by minimizing waste, promoting material recovery, and reducing pollution.

The chapter's objectives are to analyze solid waste, discuss waste management challenges in developing countries, evaluate disposal impacts, and present sustainable reduction methods. Urbanization and population growth worsen the waste crisis, requiring joint efforts in policy enforcement, technology promotion, and community involvement for cleaner outcomes.

2.2 Types and Sources of Solid Waste

Global solid waste management prioritizes classifying waste for treatment, disposal, and recycling to be effective. Waste is broadly categorized into hazardous and non-hazardous types for proper management (Rajendran *et al.*, 2021; Varjani *et al.*, 2021).

2.2.1 Classification of solid waste

Solid waste can be classified into several categories (see Table 2.1) according to its kind and source.

2.2.1.1 Municipal solid waste

Municipal solid waste (MSW) is waste resulting from the daily activities of households, offices, and public places. This includes food scraps, paper, plastic, textile, glass, etc. Management of municipal solid waste is mainly carried out by local authorities or private waste management companies.

Table 2.1. A detailed study of the major sources and types of solid waste. Author's own interpretation.

Source	Typical facilities/activities	Types of waste
Residential	Single-family homes, apartments	Food waste, paper, plastics, metals, glass, textiles, yard waste, special wastes such as batteries, oil, and tires.
Commercial	Retail stores, restaurants, offices, hotels	Paper, plastics, wood, food wastes, metals, and hazardous wastes.
Institutional	Schools, hospitals, government buildings	Similar to commercial waste, plus medical wastes and hazardous wastes.
Industrial	Factories, refineries, power generation plants	Process wastes, hazardous chemicals, scrap metals, paper, glass, and plastics.
Construction and demolition	Construction sites, road repair, renovation	Concrete, wood, steel, bricks, and rubble.
Municipal services	Street cleaning, parks, recreational areas	Street sweeping, landscaping debris, and general refuse.
Agriculture	Farms, orchards, vineyards, feedlots	Spoiled food, crop residues, hazardous agricultural waste.
Treatment facilities	Water, wastewater, industrial treatment processes, etc.	Treatment plant wastes, principally composed of residual sludges and other residual materials.

2.2.1.2   Hazardous waste

Hazardous wastes are those that can be harmful because of various factors and can be manifested in solid, liquid, and gaseous forms. Some examples include batteries, pesticides and chemical solvents, medical waste, and heavy metals containing waste from industrial activities (Khan *et al.*, 2018).

2.2.1.3   Industrial waste

Industrial wastes from factories, mills, mines, and manufacturing units can be hazardous (toxic chemicals, heavy metals) or non-hazardous (paper, plastics, glass, wood scraps).

2.2.1.4   Electronic waste

E-waste includes discarded electronics like computers, phones, televisions, and circuit boards. This sector grows due to technology advancements and short life cycles of products. It contains valuable materials (gold, silver, platinum) and harmful substances (lead, mercury, cadmium).

2.2.1.5   Biodegradable waste

These include wastes of organic nature that decompose naturally through microbial action. Examples are food waste, yard waste, and agriculture residues.

2.2.1.6   Inert waste

Inert waste consists of materials that do not decompose over time. An example would be construction and demolition debris like concrete, bricks, and rubble.

2.2.1.7   Radioactive waste

Waste originating from nuclear plants, medicine, and research poses lasting risks to the environment and health, necessitating specialized handling and disposal procedures.

2.3   Solid Waste Generation and Composition

Solid waste management is crucial for health, nature, and economy, as waste arises from various sources. Understanding waste types is essential for effective management, including organic, inorganic, and hazardous materials.

2.3.1   Factors influencing waste generation

Solid waste generation varies regionally due to factors such as population growth, urbanization, economic development, and consumption patterns, which influence the volume and composition of waste in different areas (Abir *et al.*, 2023).

- *Population growth*: drives demand for goods, leading to increased waste. Urban areas face challenges managing waste due to high population density, necessitating improved collection and disposal methods (Kumar *et al.*, 2017).
- *Urbanization*: increases solid waste production by driving higher consumption and disposal rates of packaged goods, construction materials, and industrial products in expanding city areas. This strains waste management systems meant for smaller capacities, emphasizing the urgency for improved solutions.
- *Economic development*: economic growth drives consumption, alters production, and boosts waste production. Developed economies, with high income and technology, generate industrial waste, requiring stringent waste management for both hazardous and non-hazardous waste.
- *Consumption patterns*: changing lifestyle and consumption behaviors significantly impact solid waste by increasing plastic, paper, and electronic waste in high-consumption societies. Convenience-driven consumption patterns also contribute to rising non-biodegradable waste fractions, notably from packaged food and disposable products.
- *Composition of waste*: factors like location, socioeconomics, and policies influence solid waste composition, dividing it into organic and inorganic fractions with distinct environmental implications. Organic waste, such as food scraps and yard waste, makes up half of the waste and can be composted

to reduce landfill pressure and emissions. Inorganic waste, like plastic and metal, can be recycled to address its slow decomposition. Hazardous waste, comprising heavy metals and biomedical waste, must be managed properly to protect health and the environment. Diverse waste types demand customized management to reduce risks and enhance sustainability (Abir *et al.*, 2023).

2.4 Impacts of Improper Solid Waste Management

Inadequate waste management harms ecosystems and worsens air, water, and soil quality, exacerbating global environmental issues like pollution and climate change (Abubakar *et al.*, 2022).

2.4.1 Environmental impacts

- *Air pollution*: poor waste management mainly causes air pollution through landfills, including methane emissions and open burnings. Methane from landfills worsens climate by decomposing organic waste anaerobically, releasing greenhouse gas (GHG) 28 times stronger than CO_2. Burning solid waste emits harmful gases like CO, NO_x, SO_2, and PM (particle matter), while plastic incineration releases toxic substances causing cancer and endocrine issues (Browning *et al.*, 2021). Volatile organic compounds released during degradation, including benzene and formaldehyde, contribute to ozone, smog, and health issues such as respiratory diseases and cardiovascular problems.
- *Water pollution*: improper waste disposal contaminates water sources with toxins through leachate pollution from landfills. Plastic waste breaks down into harmful microplastics, impacting ecosystems. Eutrophication and algal blooms disrupt aquatic ecosystems by increasing nutrient levels, causing oxygen depletion (Brahimi *et al.*, 2019).
- *Soil contamination*: uncontrolled dumping of solid waste harms soil fertility and agricultural productivity by causing hazardous substance build-up. This includes heavy metals like lead and cadmium from e-waste, which pollute soil and hinder plant growth, as well as persistent organic pollutants from dumped plastics, which affect plant uptake. Waste decomposition can alter soil pH, decrease productivity, and lead to acidification, vegetation changes, compaction, impaired water flow, and erosion (Usman *et al.*, 2017).
- *Greenhouse gas emissions and climate change*: solid waste management significantly impacts global greenhouse gas emissions by contributing to climate change through methane and carbon dioxide emissions from landfills, black carbon emissions from waste incineration, and deforestation caused by uncontrolled waste dumping, which reduces carbon sequestration capacity and worsens climate change (Ramachandra *et al.*, 2018).

2.4.2 Health impacts

The health effects of exposure to solid waste depend on waste type, management practices, exposed population, duration of exposure, and intervention effectiveness (Bhardwaj and Jindal, 2022). Waste-related health impacts include infectious diseases transmitted via medical waste, attracting disease vectors like mosquitoes and rodents that can spread diseases such as hepatitis, tuberculosis, and malaria to humans. Improper disposal sites can also spread zoonotic diseases, while chronic exposure to hazardous waste like heavy metals and persistent organic pollutants can cause cancer, neurotoxicity, reproductive, and cardiovascular disorders in humans (Vinti *et al.*, 2021; Khan *et al.*, 2023).

2.4.2.1 Vector-borne diseases

Improper waste disposal fosters vectors like mosquitoes, flies, and rodents, which spread pathogens causing diseases like malaria, dengue fever, and leptospirosis.

1. *Malaria and dengue fever*: malaria, Plasmodium, and Anopheles dengue vectors are the causes of the above diseases. Dengue-vector Aedes spread transmission in places with solid waste. Lack of appropriate sanitary disposal leads to breeding sites for mosquitoes and attracts flies that contaminate food and water. Stagnation of water in drains and dumps favors the transmission of diseases. Fever, chills, anemia, and organ failure are symptoms of malaria. High fever, joint pains, rashes, and bleeding are other manifestations. For severe illness, there might be dengue shock syndrome. Waste management, maintenance of drainage, and vector control through larvicides and biological methods prevent such diseases.

2. *Leptospirosis*: leptospirosis is an infectious disease caused by spirochete bacteria of the genus Leptospira. The bacteria infect the host through urine, especially from rodents, and thrive in contaminated waters mainly in flood-prone areas where waste accumulates. Symptoms arise from mild cases such as fever and muscle pains to severe conditions like renal failure, meningitis, hepatic damage, and respiratory distress.

2.4.2.2   Respiratory issues

A large part of internal pollution and health problems in India is associated with improper solid waste disposal methods. Open burning generates PM2.5, PM10, carbon monoxide, dioxins, furans, and heavy metals, which in turn may cause chronic obstructive pulmonary disease, asthma, lung cancer, and heart diseases. These toxins interfere with endocrine and immune functions. Landfills release airborne pathogens, fungal spores, and endotoxins, which cause allergies, infections, and respiratory diseases (Brahimi *et al.*, 2019; Abubakar *et al.*, 2022). Waste workers are exposed to much higher risks and should be provided with personal protective equipment (PPE), maintain ventilation, and are subject to health screening.

2.4.2.3   Exposure to toxic substances

Toxic elements such as lead, cadmium, mercury, arsenic, and chromium are found in e-waste, industrial by-products and medical wastes contain heavy metals, persistent organic pollutants (POPs), and radioactive materials, posing health risks (Bhardwaj *et al.*, 2024). Lead disrupts child development, causing cognitive and behavioral issues. Mercury and cadmium are also harmful, leading to kidney failure, skeletal deformities, and neurotoxicity. Medical waste, including expired pharmaceuticals and cytotoxic drugs, contains toxic chemicals. Improper disposal poses risks of infectious diseases. Pharmaceutical waste exposure affects endocrine function and promotes antimicrobial resistance. By enforcing strict biomedical waste handling rules, applying advanced treatment methods such as autoclaving and incineration, and using pollution control systems should lessen the effects of exposure.

2.4.2.4   Psychological stress

Improper handling of waste causes social and economic issues, especially for those in poverty. Living near dump sites brings psychological distress, such as odors and diseases, escalating anxiety, and depression. Salvagers face stigma and stress working without protective gear near hazardous waste. Health risks can be reduced by noting direct effects from dangerous materials and indirect effects like air and groundwater pollution. Methane from landfills contaminates groundwater and toxic pollutants contribute to respiratory diseases (Abubakar *et al.*, 2022).

2.4.3   Safety concerns

Proper waste management is vital for the environment and public health, yet it poses significant safety hazards. Neglected disposal sites jeopardize workers, communities, and the environment, necessitating the understanding of safety concerns for crafting protective waste policies.

1. *Occupational hazards for waste handlers*: sanitation workers, recyclers, landfill personnel, and waste handlers face various job hazards, such as contact with toxic materials or sharp objects that can cause injuries or health issues (Gowda *et al.*, 2023). There

are many hazards in the workplace, including physical injuries like cuts and fractures, chemical exposure which causes burn and organ damage, and biological exposure hazards due to pathogens causing infections. Heavy lifting and repetitive motions lead to ergonomic stress, which can cause chronic pain. Safety regulations along with PPE and regular training are included to lessen these risks and make workplaces safe.

2. *Fire and explosion risks*: the situation in dump sites, where flammable materials could be stored, increases the possibility of fire and explosion. Methane released from decomposing wastes can accumulate, igniting the gases if not released or vented. Improper disposal of chemicals, batteries, and other highly volatile materials can start fires or create toxic fumes. Burning of waste in open places increases pollution levels and can cause outbursts of uncontrollable fire. Prevention measures will include gas monitoring systems, fire safety systems, and proper disposal of hazardous waste.

3. *Structural hazards from unregulated dumpsites*: there are serious risks of structural failure when it comes to unregulated dumpsites without due management. Poorly engineered landfills can subside, damaging infrastructure and posing risks to surrounding communities. Unstable waste piles can collapse, trapping workers and scavengers. Insufficient liners and drainage allow leachate to seep into drinking water and groundwater systems. If waste disposal sites were brought under regulation, enforced engineering standards made, and routine inspections conducted, then these structural hazards could be minimized, and any impact on workers and the environment could be averted.

2.5   Solid Waste Management Strategies

Collection and transportation of waste are crucial for an efficient waste management system. Well-designed processes ensure timely and eco-friendly waste removal to prevent health hazards and pollution (Ziraba *et al.*, 2016). Recent advancements and technology have enhanced these systems, making them more sustainable and cost-effective.

2.5.1   Waste collection and transportation

1. *Collection methods* (door-to-door, community bins, automated systems): Such means of waste collection depend entirely on the level of urbanization, population density, and infrastructure present. Household collection methods can ensure better waste segregation, however, they are costly in terms of labor and operations. Community bins and drop-off points function well in areas without house-to-house collection, yet their success depends on proper maintenance and public awareness. The automated waste collection system (AWCS) employs vacuum tubes to transfer waste to central stations, thereby minimizing emissions and enhancing sanitation especially in urban areas, however, the initial establishment cost may be high.

2. *Vehicles and infrastructure*: the waste collection has special vehicles and infrastructure to transport as economically as possible. For instance, there are compactor trucks that compress wastes to lessen the volume for transportation costs, while dump and open-body trucks are used for loose bulk materials such as construction waste. A recycling truck has compartments for separating materials, and inlet vacuum trucks can collect liquid and hazardous waste. Transfer stations are areas where refuse can be temporarily stored and compacted for more cost-effective long-distance transport to a landfill. Reduced congestion and emissions are produced through transport routes solely dedicated to the collection process. Additionally, they minimize road congestion using train and water transport instead of roads. Combined with right technology in advanced vehicles, strategic collection methods, and efficient infrastructures, it tends to lend itself to sustainable waste management.

2.5.2 Waste segregation and recycling

Waste segregation and recycling support sustainable waste management by minimizing landfill usage and mitigating environmental impacts (Bui *et al.*, 2022). Efficient segregation aids recycling, but challenges such as contamination and poor infrastructure hinder effective waste management. The informal sector plays a vital role in waste recovery and material recycling, particularly in developing countries.

1. *Source segregation techniques*: waste source segregation enhances the recycling efficiency and disposal by separating the waste at source. The waste is categorized in color-coded bins – green for organic waste, blue for recyclable, red/yellow for hazardous, and black for non-recyclable materials. Dual and multi-stream separation divides the materials further for better processing. On-site composting reduces the load of waste to the municipality while enriching the soil. Extended producer responsibility (EPR) makes product design recyclable at the producers' end and grants the responsibility for the management of post-consumer waste to manufacturers.
2. *Recycling process and challenges*: the recycling process is divided into collection and sorting, during which waste is separated from households, businesses, and collection points, either manually or mechanically with the help of conveyor belts, optical sorters, and magnetic separators. Next comes the processing and cleaning of recyclables whereby they are cleaned, shredded, or melted for reuse. Re-manufacturing at the end converts recovered materials into new products such as recycled paper, plastic pellets, and metal sheets. Despite its environmental and economic advantages, recycling faces challenges in (i) contamination – an introduction of organic or hazardous waste compromises the quality of recyclables; (ii) public awareness that is lacking concerning segregation of recyclable materials; (iii) infrastructure and market concerns which arise due to limited facilities and fluctuating price of materials; and (iv) high processing costs, especially for multilayer plastics and complex materials.

3. *Role of the informal sector in recycling*: the informal sector plays a vital role in recycling, especially in low- and middle-income countries, through waste collection and sorting, where waste pickers and scrap dealers manually recover recyclables, reducing landfill waste. This sector has significant economic and environmental impacts, providing livelihoods to marginalized communities while conserving resources and reducing pollution. However, challenges such as hazardous working conditions, lack of social security, and exclusion from formal recycling networks hinder their efficiency. Recognizing and integrating informal workers into formal waste management can improve recycling rates while enhancing their working conditions and economic security.

2.5.3 Waste treatment and disposal methods

Various waste management techniques aim to reduce environmental harm and maximize resource recovery. Effective strategies ensure safe waste disposal, decrease landfill dependence, and encourage sustainable waste practices.

2.5.3.1 Composting and organic waste processing

Organic waste composting transforms food scraps and garden waste into nutrient-rich soil amendments through microbial activities. It involves processing various biodegradable materials to create valuable compost (Srivastava *et al.*, 2015).

1. *Aerobic composting*: microorganisms decay organic matter with oxygen, creating compost through aeration, moisture, and temperature control.
2. *Vermi composting*: enhances decomposition of organic waste using earthworms, thereby increasing nutrient availability in the compost product.
3. *In-vessel composting*: this is a type of controlled composting inside an enclosed

system for the purposes of speeding up the decomposition as well as odor control, characteristic of urban areas.

2.5.3.2 Thermal treatment technologies

Thermal treatment uses high temperatures for breaking down wastes, reducing volume, and recovering energy. The treatment approach depends on waste composition, land access, and cost factors. The important thermal treatment processes include:

1. *Incineration*: a method of disposing of wastes by burning the wastes at temperatures above 1000°C for maximum oxidation and reducing the volume of waste by 80–90%. Advanced plants within the range of 65–80% thermal efficiency produce electricity with stringent pollution controls (Das *et al.*, 2018; Khan *et al.*, 2023). Two types of incinerators are stoker furnaces for large-scale municipal waste incineration and fluidized bed furnaces that promote efficient heat transfer and enhanced efficiency. Advantages include energy recovery and silent, odorless operation. On the downside, the process generates dioxins, SO_x, NO_x, and particulate matter; it is also costly and cannot use regular handling. In developing countries, poor maintenance often ends up in hazardous waste mismanagement.
2. *Pyrolysis*: a type of thermal decomposition of waste in the absence of oxygen producing biochar, syngas, and bio-oil. Pyrolysis is a heat-driven conversion process for solid waste, occurring without oxygen at temperatures of 400–1000°C. It outputs syngas, char, pyrolysate oil, and a mix of gases such as CO, CH_4, and H_2. Efficiency varies with feedstock, with plastics, rubber, e-waste, and wood being optimal. Advancements have led to interest in recycling tires to reclaim oil, carbon black, wire, and gaseous fuels. It is suitable for treating plastics, tires, and biomass and facilitates energy recovery and material recycling.
3. *Gasification*: high-temperature gasification with limited oxygen converts carbon-containing waste into syngas (a mixture of hydrogen, carbon monoxide, and methane).

This makes it a good energy-efficient option for converting waste into biofuels, hydrogen, or electricity because it simultaneously decreases waste volume by almost 95%, which is best for dealing with municipal solid waste. Uses include syngas and slag use in industries, and the recovery of sulfuric acid from sulfur-containing waste. Disadvantages to gasification are that it forms toxic polyhalogenated organic compounds and an input of carbon-rich material for process efficiency.

4. *Hydrothermal carbonization*: hydrothermal carbonization (HTC) is a thermal process converting moist solid waste at 180–350°C with water to produce hydrochar, a carbonaceous residue suitable for biofuel. Benefits include low energy requirements, water-friendly for wet waste, yet costlier for large-scale implementation (Das *et al.*, 2018).

Thermochemical conversion in waste-to-energy (WTE) systems turns MSW into heat, syngas, biochar, and compost, with method choice based on waste attributes, eco-concerns, and cost. Pyrolysis and gasification, preferred for efficiency and eco-friendliness, surpass incineration. Eco-friendly waste management includes composting and carbonization.

2.5.3.3 Sanitary landfilling

Landfilling remains mostly used for non-recycling or inert wastes, despite advances in waste treatment, because of the cost and low labor requirement of landfilling. Sanitary landfills reduce the potential negative environmental and health impacts that might arise through their engineered systems (Nanda and Berruti, 2021). Synthetic/clay liners and leachate treatment prevent groundwater contamination from leachate through liner and leachate management. Collection systems collect the methane gas to use it for energy or flaring in reducing emissions. Odor, pests, and windblown dispersal are often controlled through compaction and covering. The necessity of landfilling for some types of waste is well known; however, in current waste practices, waste reduction and separation and alternative treatment methods are preferable. Regulations vary: while developed countries impose strict control on leachates and methane, countries like Pakistan tend to have it in open

Table 2.2. Categories of solid disposal. From Khan *et al.* (2023). Table used under license CC-BY-4.0.

Category	Types of solid waste disposal
Class 01	Disposal of soil and excavated materials.
Class 02	Disposal of construction, demolition, and renovation waste, including mineral-based materials.
Class 03	Disposal of municipal solid waste, including household and commercial refuse.
Class 04	Disposal of industrial and commercial waste, excluding hazardous substances.
Class 05	Disposal of hazardous waste, including toxic, chemical, and biomedical waste.
Class 06	Underground disposal of specialized waste, such as radioactive materials

waste disposal at risk to health and livelihood (Nanda and Berruti, 2021). Solid waste can be classified based on its origin and composition as shown in Table 2.2 to ensure the most efficient waste disposal methods.

1. Bioreactor landfills are modern MSW management methods shifting from waste disposal to active treatment. They accelerate organic waste decomposition and stabilization via controlled biological techniques. Advantages include improved leachate quality, reduced reliance on external treatment, increased landfill gas production with higher methane yields, cost-effective energy production, and faster waste stabilization, settlement, and maintenance cost reduction. Internal temperature ranges from 45 to 60°C, optimal for methanogenic bacteria growth at 35 to 45°C in an alkaline environment (pH 6–8, alkalinity of 2000 mg/L). Bioreactor landfills are categorized by the presence of oxygen and microbial activity: anaerobic, aerobic, and semi-aerobic types.

2. Anaerobic bioreactor landfills use anaerobic bacteria to convert organic matter into landfill gases like methane and carbon dioxide. This process includes hydrolysis, acidogenesis, acetogenesis, methanogenesis, and fermentation. Methane produced can be used for renewable energy.

3. Aerobic bioreactor landfills use oxygen to break down organic waste quickly, reducing methane emissions, improving leachate quality, and stabilizing the landfill faster. By promoting aerobic conditions, these landfills efficiently degrade pollutants, with methane concentrations dropping from 60 to 10% or less in 7 to 10 days, suppressing methane emissions.

4. Semi-aerobic bioreactor landfills maintain partially oxygenated conditions to support aerobic and anaerobic microbial communities together, allowing for simultaneous methanogenesis, hydrolysis, and fermentation at different oxygen levels. Key features include air circulation for dual degradation, diverse microbes for efficient waste breakdown, and leachate collection systems for passive aeration, reducing external aeration needs. Unlike fully aerobic bioreactors, semi-aerobic landfills use natural convection through leachate pipes to create oxygen gradients, promoting effective aerobic and anaerobic decomposition in the waste mass (Nanda and Berruti, 2021).

2.5.4 Emerging technologies in waste treatment

Resource recovery and mitigation of environmental impacts have instigated the development of innovative technologies for waste treatment (Armah *et al.*, 2020). Some prominent emerging technologies are:

1. *Plasma arc gasification*: employs high-temperature plasma torches (between 3000 and 40,000°C) to reduce waste into its elemental components. Produces syngas and slag in a glass-like form with minimum emissions and high energy efficiency. Suitable for treating hazardous and mixed waste streams.

2. *Bio-digestion* (anaerobic digestion): decompose organic waste by microorganisms in an anaerobic condition producing biogas and producible digestate-microbial fertilizer. Such a system is of great relevance for managing food waste, agricultural residues, and sludge. Biogas fits the description of a renewable energy source that is convertible into electricity, heat, or vehicle fuel.

2.6   Policy and Legislative Framework for Waste Management

Effective waste management relies on robust policy and regulations safeguarding the environment, public health, and resources. National and global waste policies establish legal structures. Governments and cities enforce these practices, while frameworks like the extended producer responsibility promote eco-friendly strategies (Ray and Rahman, 2016).

2.6.1   National and international waste management policies

Waste management policies aim to limit society's waste generation, handling disposal and resource recovery. They vary by country and may involve treaties, laws, and best practices. Table 2.3 highlights key legal milestones in India's journey toward effective waste management.

Table 2.3. Important legal turning points in the history of waste management in India. From Ray and Rahman (2016). Table used under license CC-BY-3.0.

Rules/acts/criminal law	Years	Salient features
Indian Penal Code, 1860	1860	Solid waste management, treated as a public nuisance under chapter XIV, addresses offenses affecting public health, safety, convenience, decency, and morals with associated penalties.
Criminal Procedure Code, 1973	1973	Section 133 empowers magistrates to remove public nuisances and stop offending businesses.
The Water (Prevention and Control of Pollution) Act, 1974	1974	Empowers boards to prevent water pollution, monitor quality, and penalize violators.
The Water (Prevention and Control of Pollution) Cess Act, 1977	1977	Provisions introduced for levying cess on water used in landfilling, composting, and digesters.
Hazardous Wastes (Management and Handling) Rules, 1989	1989	Identifies 44 waste-generating processes and outlines disposal guidelines.
The Biomedical Waste (Management and Handling) Rules, 1998	1998	Requires healthcare institutions to manage hospital waste through proper segregation, collection, disposal, and treatment.
Municipal Solid Wastes (Management and Handling) Rules, 2000	2000	Municipal authorities manage solid waste, including door-to-door collection, segregation, street cleaning, closed truck transportation, composting, waste-to-energy plants, and landfill standards for disposal.
The Batteries (Management and Handling) Rules, 2001	2001	Regulates safe disposal of batteries for various parties to ensure environmental safety.
Plastic Waste Rules, 2011	2011	Requires a minimum of 40 microns for plastic bags, bans free distribution, and limits recycled/compostable plastics for food.
E-waste (Management and Handling) Rules, 2011	2011	Defines e-waste terms like authorization, bulk consumer, historical e-waste, sound management, and recycling. Mandates EPR, requiring producers to handle end-of-life e-waste.

1. *National waste management policies*: in terms of national waste management policies, the approaches that govern everything from waste reduction to recycling, treatment, and disposal are the key ones. The waste hierarchy that provides priority to prevention, minimization, reuse, recycling, waste-to-energy, and safe disposal stands as the prime guideline. The legal framework deals with segregation, hazardous waste handling, and landfill management. Financial instruments are placed on waste disposal fees, landfill taxes, and disincentives to recycling industries. The citizen engagement level has been achieved through public education and awareness campaigns held under the umbrella of the Resource Conservation and Recovery Act (USA), Waste Framework Directive (EU), and Solid Waste Management Rules (India) that persuade citizens to adopt responsible waste management practices.

2. *International waste management policies*: international agreements play a significant role in a country's waste management policies, especially with regards to hazardous materials. The Basel Convention from 1989 prohibits the transboundary movement of hazardous waste across member states to prevent illegal dumping. The Stockholm Convention of 2001 regulates the production, use, and disposal of persistent organic pollutants (POPs), including industrial chemicals and pesticides. The Minamata Convention of 2013 seeks, among other objectives, to minimize the pollution of mercury from waste and the various industries responsible for its creation. The UN Sustainable Development Goal (SDG) 12 on responsible consumption and production urges the sustainability of waste management. The above international legal conditions help create a partnership for international cooperation on the sustainable practices related to waste.

2.6.2 Role of government and municipal bodies

Waste management policies are being formulated, implemented, and monitored by governments and municipal authorities. Their responsibilities include:

1. *Functions of national government*: the national governments are crucial in waste management through the following major functions: legislation and regulation, both by enacting and amending laws on waste management; providing funds and developing the infrastructures to ensure that resources are allocated for facilities to treat wastes, recycle wastes, and manage landfills; and monitoring and compliance with the enforcement of waste laws under regulatory agencies, including penalties for violations. Most innovative reductions in waste, treatment methodologies, and practices in the circular economy are fostered through research and development aimed at sustainable waste management.

2. *Functions of municipal and local government*: the responsibility for the collection and transport of waste falls on municipal and local authorities and these bodies seek both promptness and proper disposal. They introduce various segregation and recycling programs by establishing material recovery and composting facilities. Awareness campaigns are held year-round to educate the communities on waste reduction, segregation, and disposal. This awareness is also extended to coordination with the private sector and informal workers by integrating waste collectors into the formal management system thereby promoting efficiency and sustainability.

2.6.3 Waste management regulations and guidelines

The subject of regulations and guidelines for waste management encompasses principles and terms for the operation of many processes governing waste disposal methods. Solid waste regulations are concerned with municipal waste collection, transportation, and disposal. Hazardous waste management rules apply to any activity involving hazardous materials in handling, treatment, or disposal. E-waste regulations call for the responsible recycling of products with hazardous materials. Landfill regulations set design standards and those for leachate and gas collection systems to minimize damage to the environment. Plastic waste management regulations restrict single-use plastics and encourage alternatives that are biodegradable. These regulations, when complied with, are responsible for environmental safety and efficient waste management.

2.6.4   Extended producer responsibility

Extended producer responsibility (EPR) promotes waste reduction and recycling by ensuring that manufacturers take care of their products throughout their entire life cycles – from design to post-consumer disposal. The most important features of EPR are environmentally friendly design, take-back programs for collection of products at the end of their useful life, and producer financing of waste management systems for recycling. Regulation applies in various industries such as the electronics, packaging, and automotive sectors. EPR applications reduce packaging waste, enhance recycling, reduce municipal solid waste costs, and protect the environment. Countries with EPR models like Germany, Canada, and Japan have been successful in proving its significance in sustainability and waste reduction.

2.7   Sustainable and Innovative Waste Management Practices

Traditional waste management faces modern global challenges; sustainable methods offer new solutions, emphasizing integrated solid waste management (ISWM) techniques, waste hierarchy, 3Rs, and life cycle thinking worldwide. Developing economies lack effective reuse and recycling methods, leading to un-recycled waste and environmental issues. Inadequate waste disposal contributes to pollution and health risks. Efficient MSW management depends on enforcement, resources, waste characteristics, and proper collection. Solid waste complexities hinder segregation and recycling, requiring systematic solutions like source separation and

enhanced infrastructure. Innovative practices such as artificial intelligence (AI) and the Internet of Things (IoT) applications enhance waste management strategies globally.

2.7.1   Circular economy approach

The circular economy minimizes waste and maximizes resource efficiency by ensuring that materials remain continuously available in use through sustainable design, recycling, and/or reuse. The major principles are:

- Sustainable product design toward longevity, reparability, and recyclability.
- Recovering materials to avoid raw material extraction through reuse such as the reuse of metals, plastics, and organic waste.
- Waste as resource is the use of discarded materials into new products.
- Provide an infrastructure for industrial symbiosis in which industries use by-products of other industries as raw materials.
- Deposit-refund schemes encourage product returns for recycling (e.g. bottles, electronics).
- Eco-industrial parks facilitate resource-sharing among industries.

2.7.2   Zero waste strategies

Zero waste strategies are described as preventing waste, realizing sustainable consumption, and recovering resources from waste to lessen the environmental footprint (Das *et al.*, 2019; Awasthi *et al.*, 2021). Fundamental elements include:

- Waste prevention via sustainable design together with limited disposable goods.
- Product stewardship in which manufacturers are held accountable for the whole life cycle of a product.
- Extended producer responsibility (EPR) to enhance the reusability and recyclability of product designs.
- Community engagement in which individuals and businesses are educated on sustainable consumption.

Advantages which include:

- Environmental: it is witnessed as reduction in pollution, resource extractions, and greenhouse gas emissions.
- Economic: the generation of employments in recycling, composting, and green manufacturing.
- Public health: it reduces toxics exposure from landfills and incineration.

2.7.3 Technologies for waste to energy

Waste-to-energy technologies (WTE) convert wastes into energy or fuel, including: electricity, heat and biofuels, and partially empty the landing sites and creating renewable energies.

- Incineration: the burning of wastes at high temperatures to produce steam to drive turbines in electricity generation and it has modern pollution control equipment.
- Gasification: this is the conversion of carbon-based wastes into a synthesized gas (H_2 and CO) for energy production or chemical feedstock.
- Pyrolysis refers to the heating of waste material in the absence of air, producing bio-oil, syngas, and biochar as feedstock or fuel for the industry.
- Anaerobic digestion: represents the decomposition of organic waste by micro-organisms to produce the following – biogas (mostly methane) and a nutrient-rich digestate – with the latter.

Advantages:

- Reduces methane emissions from landfills.
- Generate renewable energy; it reduced resource dependence on fossil fuel resource.
- Volume reduced up to 90% which, therefore, required much less space in landfills because the volume of waste is reduced substantially.

2.7.4 The role of AI and IoT in the waste management process

AI and IoT promotes efficiency improvements in waste management, minimizes operational costs, and optimizes resource utilization.

2.7.4.1 The role of AI in waste management

- Automated waste sorting: with robots and algorithms to aid correct sorting, useful recycling can therefore be enhanced by artificial intelligence.
- Predictive analysis: to reduce transportation cost, gaps in collection schedules can be analyzed by AI models and by properly analyzing waste generation schedules.

- Smart recycling bins: AI-enabled bins suggest the waste type and fills the level of waste in real time for collection efficiency.

2.7.4.2 IoT-based waste management

- Smart waste bins: embedded sensors in smart waste bins measure the level of waste and inform collectors, thereby curtailing unnecessary pickups.
- Fleet management systems: these systems use GPS and IoT sensors to optimize waste collection routes by decreasing fuel consumption and emissions.
- Through waste tracking and monitoring, real-time data analytics allow municipalities and businesses to account for waste streams and regulatory compliance.
- Sustainable waste management through innovative propositions combining AI and IoT assures environmental sustainability and promotes resource efficiency.

2.8 Public Awareness and Community Participation

It is in the interest of the community that active public participation is the backbone for sustainable waste management. It involves educational campaigns, citizen involvement, and corporate responsibility as just a few shades in promoting a culture of proper waste management, recycling, and environmental participation.

2.8.1   Role of educational campaigns

Educating schools, households, businesses, and policymakers on waste reduction and appropriate means of disposal is encompassed in these campaigns. Examples of strategies include school initiatives, community workshops, gamified incentives, and media campaigns, all of which are in line with the UN Sustainable Lifestyles programs that produce long-term behavior change.

2.8.2   Citizens' participation in recycling and waste segregation

Public participation is the principal element in good waste management, assuring better waste segregation, little pollution, and less waste at landfills. Modes of engaging the public include:

*   Education initiatives by volunteers and agencies.
*   Municipal regulations enforcing waste segregation.
*   Color-coded bins to facilitate proper sorting.
*   Recycling centers and mobile apps for waste tracking.

San Francisco and Singapore have successful models where citizen-driven recycling decreases landfill dependency.

2.8.3   Corporate social responsibility initiatives

Corporations are important agents for the purpose of waste reduction via sustainable business practices, recycling programs, and product return systems. They contribute through:

*   Reducing packaging wastes and using recycled materials.
*   Partnership with waste management agencies for infrastructure development.
*   Involvement of employees in waste reduction programs.
*   Funding for research and environmental projects.

Recognition of such major corporate social responsibility (CSR) programs includes:

*   Unilever's Sustainable Living Plan (waste reduction and recycled plastics).
*   Coca-Cola's 'World Without Waste' (100% bottle recycling by 2030).
*   Apple's Trade-in and Recycling Program (e-waste reduction).
*   Nestlé's commitment to plastic-free packaging by 2025.

CSR initiatives cut across SDGs making the allocation more beneficial to positive corporate branding and sustainability.

2.8.4   Challenges in the public awareness and participation

Some of the barriers to effective participation may include:

*   Lack of awareness on waste segregation and recycling.
*   Deep-rooted disposal habits resistant to change.
*   Minimum facilities for waste collection and recycling.
*   Weak enforcement of regulations and socioeconomic constraints.
*   Stronger policies, public education, and improved waste management infrastructure will suffice to tackle these issues in the bid for long-term community involvement.

2.9   Case Studies and Best Practice

Waste management has very much to do with the presence of regulatory frameworks, public participation, technology, and financial investment subsequently. Table 2.4 presents a comprehensive systematic review on studies involving an analysis of the effects of these multiple factors on solid waste management systems across Asia. Successful models are good guides for this.

Table 2.4. Literature discussing the impact of solid waste management in Asia. From Abubakar et al. (2022), courtesy of the National Library of Medicine.

Authors	Study area	Study aim	Impact on humans	Impacts on the environment	Recommendations/implications
Ossama et al. (2020)	Saudi Arabia	Reviews municipal SWM practices	Infections from landfill gases and leachate	Air pollution, water contamination	Recycling and reducing landfill reliance
Akmal and Jamil, 2021	Rawalpindi and Islamabad, Pakistan	Explores overall the relationship between dumpsite exposure and the health of residents	Malaria, dengue fever, respiratory diseases, skin diseases, diarrhea due to dumpsites	Groundwater contamination, land pollution from waste disposal in drains, roads, and railway tracks	Improve waste-to-energy (WTE) industry, remove illegal dumpsites, and public awareness campaigns
Hoang and Fogarassy, 2020	Hanoi, Vietnam	Explores sustainable MSW management options	Public health threats from GHG emissions, strong odors	Land and soil degradation, air and water pollution	Mechanical-biological treatment (MBT) for sustainability
Brahimi et al. (2019)	India	Explores waste-to-energy potential	Infections from contaminated water, respiratory diseases from incineration	Air, soil, noise, and water pollution, global warming from open burning	Promote waste-to-energy solutions
Pokhrel and Viraraghavan (2005)	Kathmandu Valley, Nepal	Evaluates SWM practices	Odor and health risks from haphazard waste disposal	Polluted riverbanks and water bodies, impact on tourism	Composting, ban indiscriminate disposal
Das et al. (2018)	Kathmandu Valley, Nepal	Estimates MSW burning in five municipalities	Respiratory infections, allergic reactions, heart diseases	CO_2 and CH_4 emissions contribute to global warming	Improve waste segregation, penalize open burning
Usman et al. (2017)	Faisalabad, Pakistan	Investigates impact of open dumping on groundwater	Odor, visual impact, health risks	CO_2 and CH_4 emissions, soil degradation, surface/ groundwater contamination	Effective monitoring and supervision for waste disposal
Clarke et al. (2017)	Qatar	Collects resident opinions on solid waste strategies	Health risks from landfill emissions	Plastic littering, garbage on beaches, water pollution	Promote behavioral change for sustainable SWM
De and Debnath (2016)	Kolkata, India	Investigates health effects of solid waste disposal	Malaria, dengue, diarrhea from open dumping	Water and air pollution from indiscriminate waste disposal	Establish proper MSW dumping sites
Islam (2016)	Dhaka, Bangladesh	Develops effective SWM and recycling	Health threats from emissions, aesthetic nuisance from odor	CO_2 and CH_4 emissions pollute water bodies	Strict waste disposal rules and awareness campaigns

2.9.1 Successful waste models

1. *Sweden* – waste-to-energy leader: Sweden converts 99% of household waste into energy through advanced waste-to-energy systems, which also curtail the use of landfills. To that end, the most important elements are efficient incineration, extensive recycling programs, and public awareness campaigns. There is even an importing of wastes for energy production purposes. That is how far regulation frameworks can go combined with technology in waste solutions.

2. *Japan* – Kamikatsu, a town of zero waste: Kamikatsu attained this through thorough waste classification, waste exchanging centers manned or run by communities, and an ambitious goal of no landfill by 2030. It is such a powerful model of what community leverage and decentralized waste management can achieve.

3. *Germany* – EPR and the dual system: the country's extended producer responsibility (EPR) subjects producers to the liability for packaging waste. Under the Green Dot system, private companies can be granted the license to manage recycling, thus enhancing sustainability. The salient points in this are producer liability, very strict legislation, and public–private partnerships.

2.9.2 Lessons from waste management failures

Several countries excel in waste management, while others struggle with environmental degradation and health hazards. Analyzing failures can reveal key pitfalls and solutions to rectify the situation.

1. *India*: unregulated landfilling and poor waste segregation pose environmental and health risks due to millions of tons of waste dumped annually. Challenges include open dumping like Ghazipur in Delhi, weak segregation, and poor law enforcement despite regulations. Solutions involve strict enforcement of waste segregation for high recycling rates, investment in technology for sustainable waste management, and increased public awareness for responsible waste disposal.

2. *Indonesia*: marine pollution in Indonesia stems from poor waste management, especially due to inadequate infrastructure, high plastic consumption, and a lack of recycling facilities. Key solutions include expanding waste collection networks, reducing single-use plastics through policies, and enhancing recycling with private sector involvement.

2.9.3 Best practices for effective waste management

Successful implementation of waste management strategies integrates further strategies for sustainability. Some of the best practices include:

• Integrated waste management systems: this is the combination of waste reduction, recycling, composting, and energy recovery for optimum resource use and efficiency.

• Public awareness and participation: encourage community engagement and education to fortify waste sorting initiatives with incentives for waste segregation.

• Strict policy enforcement: adopt strong regulations with monitoring and compliance related to the disposal of waste.

• Advanced waste processing technology: invest in pyrolysis, gasification, AI-based waste sorting, and so forth, to ensure efficient working.

• Public–private partnership: cooperative efforts between governments, private sectors, and civil society in improving waste management services.

• Product design for sustainability: source reduction through products that are durable, repairable, and recyclable.

Global waste management methods in Sweden, Japan, and Germany show success, while India and Indonesia face challenges. Governments and communities can improve waste management for a zero-waste future.

2.10 Conclusion

To improve global waste management, adopt integrated approaches like those in Sweden, Germany, and Japan. Enhance awareness, technology, and regulations. Focus on circular models, zero waste, energy innovations, smart systems, and localized processing. Strengthen policies and collaboration. Recommendations are for policymakers to enforce rules, encourage EPR, and integrate waste management with climate policies. Industries practice sustainability and invest in green tech. Communities segregate waste, promote recycling, and advocate for better policies. Collaboration between governments, industries, and communities is crucial for sustainable waste solutions. Implementing reforms, technology, and public involvement can create a cleaner, more sustainable world.

References

Abir, T.M., Datta, M. and Saha, S.R. (2023) Assessing the factors influencing effective municipal solid waste management system in Barishal metropolitan areas. *Journal of Geoscience and Environment Protection* 11, 49–66. DOI: 10.4236/gep.2023.111004.

Abubakar, I.R., Maniruzzaman, K.M., Dano, U.L., AlShihri, F.S., AlShammari, M.S. *et al.* (2022) Environmental sustainability impacts of solid waste management practices in the global south. *International Journal of Environmental Research and Public Health* 19, 12717.

Akmal, T. and Jamil, F. (2021) Health impact of solid waste management practices on household: The case of metropolitans of Islamabad-Rawalpindi, Pakistan. *Heliyon* 7, e07327.

Arafat, H.A., Jijakli, K. and Ahsan, A. (2015) Environmental performance and energy recovery potential of five processes for municipal solid waste treatment. *Journal of Cleaner Production* 105, 233–240.

Armah, E.K., Chetty, M., Adedeji, J.A., Kukwa, D.T., Mutsvene, B, *et al.* (2020) Emerging trends in wastewater treatment technologies: the current perspective. In: Moujdin, I.A. and Summers, J.K. (eds) *Promising Techniques for Wastewater Treatment and Water Quality Assessment*. Intech Open. London, pp. 71–98.

Awasthi, A.K., Cheela, V.R.S., D'Adamo, I., Iacovidou, E., Islam, M.R. *et al.* (2021) Zero waste approach towards a sustainable waste management. *Resources, Environment and Sustainability* 3, 100014. DOI: 10.1016/j.resenv.2021.100014.

Bhardwaj, L. and Jindal, T. (2022) Polar ecotoxicology: Sources and toxic effects of pollutants. In: Jindal, T. (ed.) *New Frontiers in Environmental Toxicology*. Springer, Cham, Switzerland, pp. 9–14.

Bhardwaj, L., Kumar, D. and Kumar, A. (2024) Phytoremediation potential of *Ocimum sanctum*: A sustainable approach for remediation of heavy metals. In: Kumar, A., Bauddh, K. and Srivastava, S. (eds) *Phytoremediation Potential of Medicinal and Aromatic Plants*. CRC Press, Boca Raton, FL, pp. 46–57.

Brahimi, T., Kumar, C.R.J., Mohamed, A. and Alyamani, N. (2019) Sustainable waste management through waste to energy technologies in Saudi Arabia: Opportunities and environmental impacts. In: *Proceedings of the International Conference on Industrial Engineering and Operations Management.*

Browning, S., Beymer-Farris, B. and Seay, J.R. (2021) Addressing the challenges associated with plastic waste disposal and management in developing countries. *Current Opinion in Chemical Engineering* 32, 100682.

Bui, T.D., Tseng, J.W., Tseng, M.L. and Lim, M.K. (2022) Opportunities and challenges for solid waste reuse and recycling in emerging economies: A hybrid analysis. *Resources, Conservation and Recycling* 177, 105968.

Clarke, S.F., Nawaz, W., Skelhorn, C. and Amato, A. (2017) Towards a more sustainable waste management in Qatar: Retrofitting mindsets and changing behaviours. *Qscience Connect* 4.

Das, B., Bhave, P.V., Sapkota, A. and Byanju, R.M. (2018) Estimating emissions from open burning of municipal solid waste in municipalities of Nepal. *Waste Management* 79, 481–490.

Das, S., Lee, S.H., Kumar, P., Kim, K.H., Lee, S.S. *et al.* (2019) Solid waste management: Scope and the challenge of sustainability. *Journal of Cleaner Production* 228, 658–678.

De, S. and Debnath, B. (2016) Prevalence of health hazards associated with solid waste disposal – A case study of Kolkata, India. *Procedia Environmental Science* 35, 201–208.

Gowda, B., Gurusiddappa, L.H. and Kalikeri, S. (2023) Study on occupational health hazards of municipal solid waste workers – A review. *World Journal of Environmental Biosciences* 12(1), 24–31. DOI: 10.51847/dmEF1XWBtq.

Hoang, N.H. and Fogarassy, C. (2020) Sustainability evaluation of municipal solid waste management system for Hanoi (Vietnam) – why to choose the 'waste-to-energy' concept. *Sustainability* 12, 1085.

Islam, F.S. (2016) Solid waste management system in Dhaka city of Bangladesh. *Journal of Modern Science and Technology* 4, 192–209.

Kaza, S., Yao, L., Bhada-Tata, P. and Woerden, F. (2018) *What A Waste 2.0: A Global Snapshot of Solid Waste Management to 2050*. World Bank Publications, Washington, DC.

Khan, N.H., Nafees, M., Saeed, T. and Khan, A. (2018) Industrial symbiosis and industrial waste management in wood-based industries. *Journal of Industrial Pollution Control* 34, 2152–2158.

Khan, N.H., Naz, N., Nafees, M., Gul, N. and Saeed, T. (2023) Solid waste management. In: Li, P. (ed.) *Solid Waste Management: Recent Advances, New Trends and Applications. IntechOpen*. London, UK, pp. 15–28.

Kumar, S., Smith, S.R., Fowler, G., Velis, C., Kumar, S.J. *et al.* (2017) Challenges and opportunities associated with waste management in India. *Royal Society Open Science* 4, 160764.

Lino, F.A., Ismail, K.A. and Castañeda-Ayarza, J.A. (2023) Municipal solid waste treatment in Brazil: A comprehensive review. *Energy Nexus* 11, 100232.

Mohanty, S., Mishra, S. and Mohanty, A. (2022) Municipality solid waste management: A case study of smart city Bhubaneswar, Odisha. *Journal of Environmental Management and Tourism* 13, 1361–1373.

Nanda, S. and Berruti, F. (2021) Municipal solid waste management and landfilling technologies: A review. *Environmental Chemistry Letters* 19, 1433–1456.

Ossama, L., Manaf, L.B.A., Sharaai, A.H.B. and Mohamad, S.S.B. (2020) A review of municipal solid waste management practices in Saudi Arabia. *Journal of Waste Management and Disposal* 3, 209.

Pokhrel, D. and Viraraghavan, T. (2005) Municipal solid waste management in Nepal: Practices and challenges. *Waste Management* 25, 555–562.

Rajendran, N., Gurunathan, B., Han, J., Krishna, S., Ananth, A. *et al.* (2021) Recent advances in valorization of organic municipal waste into energy using biorefinery approach, environment and economic analysis. *Bioresource Technology* 337, 125498.

Ramachandra, T.V., Bharath, H.A., Kulkarni, G. and Han, S.S. (2018) Municipal solid waste: Generation, composition and GHG emissions in Bangalore, India. *Renewable and Sustainable Energy Reviews* 82, 1122–1136.

Ray, M. and Rahman, M. (2016) An overview of legal framework for waste management system in India with special allusion to SWM rules, 2016. *International Journal of Interdisciplinary and Multidisciplinary Studies (IJIMS)* 4(1), 13–19.

Srivastava, V., Ismail, S.A., Singh, P. and Singh, R.P. (2015) Urban solid waste management in the developing world with emphasis on India: Challenges and opportunities. *Reviews in Environmental Science and Bio/Technology* 14, 317–337.

Tayeh, H.N.A., Azaizeh, H. and Gerchman, Y. (2020) Circular economy in olive oil production – olive mill solid waste to ethanol and heavy metal sorbent using microwave pretreatment. *Waste Management* 113, 321–328.

Usman, M., Yasin, H., Nasir, D.A. and Mehmood, W. (2017) A case study of groundwater contamination due to open dumping of municipal solid waste in Faisalabad, Pakistan. *Earth Science Pakistan* 1, 15–16.

Varjani, S., Sha, A.V., Vyas, S. and Srivastava, V.K. (2021) Processes and prospects on valorizing solid waste for the production of valuable products employing bio-routes: A systematic review. *Chemosphere* 282, 130954.

Vinti, G., Bauza, V., Clasen, T., Medlicott, K., Tudor, T. *et al.* (2021) Municipal solid waste management and adverse health outcomes: A systematic review. *International Journal of Environmental Research and Public Health* 18, 4331.

Wu, F., Liu, X., Qu, G. and Ning, P. (2022) A critical review on extraction of valuable metals from solid waste. *Separation and Purification Technology* 122043.

Ziraba, A.K., Haregu, T.N. and Mberu, B. (2016) A review and framework for understanding the potential impact of poor solid waste management on health in developing countries. *Archives of Public Health* 74(1), 1–11.

3 Enhancing Self-Efficacy in the Hospitality Industry: The Role of Smart Technology, Operational Efficiency, and Service Innovation

Sarita Peddi[1], Geetha Manoharan[1]*, Sanjeev Kumar[2] and Mohammad Badruddoza Talukder[3]

[1]SR University, India; [2]Lovely Professional University, India; [3]International University of Business Agriculture and Technology, Bangladesh

Abstract

The rapid development of the hospitality industry, led by smart technology and service innovation, has a significant impact on employee self-efficacy, which further impacts client experiences. This chapter explores how technological innovations such as artificial intelligence (AI) and data analytics empower employees, enhancing operational efficiency and building better client relationships. These technologies enhance workforce confidence and abilities by providing tools for streamlining processes. In addition, the chapter highlights the importance of sustainability in the hospitality industry, and shows how this commitment to environmental stewardship boosts employee purpose, job satisfaction, and engagement. However, challenges persist, including the unwillingness to change and the need for extensive training to fill skill gaps. The suggestion is to institute high training programs aimed at digital literacy and to constantly provide feedback mechanisms to promote an innovative culture. In conclusion, the chapter will claim that emphasis on workplace self-efficacy is crucial to enhance performance and service quality, thus having better visitor experiences and long-term success for hospitality businesses.

3.1 Introduction

Global economies depend on the hospitality industry, which provides customer service and experience in accommodation, food and beverage, travel, and tourism. New technologies and consumer tastes are transforming this environment, with employee self-efficacy becoming a crucial component in service quality and operational performance (Gupta *et al.*, 2023).

3.1.1 Hospitality industry overview

The Indian Ministry of Tourism predicted in 2021 that India's hospitality business was expected to reach USD500 billion by 2024. Growth has increased competition, forcing firms to improve service and customer experience (Kumari *et al.*, 2024). Globalization and new digital platforms have elevated client expectations for personalized, efficient services. Thus, hotel companies use technology to improve

*Corresponding author: geethamanoharan1988@gmail.com

© CAB International 2025. *Municipal Waste Management: Policies and Strategies* (eds S. Kumar *et al.*)
DOI: 10.1079/9781836990666.0003

operations and obtain client data (Singh and Sihag, 2024). This evolution requires a professional, confident workforce that can adapt to new technology and service models.

3.1.2   Employee self-efficacy importance

Self-efficacy is the belief that one can succeed, and it is essential for workplace performance (Bandura, 1978). Hotel employees with more self-efficacy perform better, are more creative, and engage better with customers (Farooq et al., 2022). Confident personnel are more inclined to act, solve problems, and deliver excellent service, which boosts guest satisfaction. Self-efficacy promotes resilience and adaptability (Aggarwal et al., 2024), which are essential in a business with changing consumer expectations and operations. Progressive hospitality organizations implement training programs to improve self-efficacy, boost performance, improve workplace culture, and reduce staff turnover (Joshi and Gupta, 2021).

3.1.3   Smart technology and hospitality transformation

Artificial intelligence (AI), data analytics, and the Internet of Things (IoT) are improving hotel operations and client experiences (Manoharan et al., 2024a). Staff can focus on emotional intelligence-intensive issues while AI chatbots answer simple questions. This technology lets organizations examine massive consumer data to customize services and guest experiences (Sharma and Akram, 2024). As employees master these techniques, their confidence and job satisfaction rise, improving service quality (Jena and Nayak, 2023).

3.1.4   Chapter purpose and structure

This chapter examines self-efficacy, smart technology, and hospitality operational efficiency (Nisar et al., 2024). The role of smart technology in fostering personal competencies will be analyzed using the self-efficacy theoretical

framework (Farooq et al., 2022). After that, topics of self-efficacy and industrial sustainability are covered, coupled with operational efficiency. Since the development of employee self-efficacy is the only source for competitive advantage, employee development will be necessary to ensure long-term success in a business that has been constantly in a state of change.

3.2   A Theoretical Framework

This section discusses self-efficacy, including its relevance in the hospitality industry and how it influences performance (Wang et al., 2024). To understand why self-efficacy holds great importance in staff performance as well as service quality in the hospitality industry, one first needs to have a good understanding of the concepts and ideas that underpin them.

3.2.1   Self-efficacy definition

The belief in the ability to perform specific tasks is termed as self-efficacy (Bandura, 1978). Bandura's social cognition theory has described that self-beliefs influence behavior more and are a better predictor of effort and persistence than ability (Schunk, 2012). Individuals with high self-efficacy possess improved problem-solving abilities, greater resilience to failure, and more motivation to attempt difficult tasks (Ghosh et al., 2024a).

3.2.1.1   Self-efficacy theories and models

Albert Bandura, in his self-efficacy theory, identified four main causes of self-efficacy credence:

1. *Mastery experiences*: success boosts self-efficacy, whereas failures lower it. Hospitality workers who excel at customer service and conflict resolution feel more confident (Nair and Senthil Kumar, 2024).
2. *Vicarious experiences*: watching others succeed can boost confidence. By observing a professional colleague handle customer problems, others may follow suit (Seal and Gupta, 2024).

3. *Verbal persuasion*: supervisors and peers must encourage and give feedback to increase self-efficacy. Management recognition for outstanding service builds organizational confidence (Mukherjee *et al.*, 2023).

4. *Emotional and physiological states*: a person's emotional state during an activity can influence self-efficacy. High stress or anxiety may diminish confidence, while a positive emotional state can enhance it (Hossain *et al.*, 2023).

These sources influence self-efficacy and guide organizational strategies to boost it.

3.2.2 Self-efficacy and performance

Research shows self-efficacy strongly affects performance across sectors. Self-efficacy affects employee motivation and performance in hospitality, where service is key (Jauhari *et al.*, 2024). High self-efficacy employees are more likely to be proactive, use inventive problem-solving methods (Behera *et al.*, 2024), and overcome hurdles. They achieve high performance by setting hard goals and staying committed (Ghosh and Jhamb, 2021). Hospitality workers with self-efficacy provide excellent service and handle consumer queries. Self-efficacy increases job satisfaction, engagement, and customer satisfaction and loyalty (Gupta *et al.*, 2023). Thus, self-efficacy can assist hospitality and other service industries. Self-efficacy boosts employee resilience, making transitions and pressures easier to handle (Jena and Nayak, 2023). Employees in the fast-changing hospitality business need great self-efficacy to manage consumer preferences and operational issues.

3.2.3 Relevance of self-efficacy in the hospitality context

Hospitality workers need self-efficacy because of human connection. Service quality and guest experience are linked (Mukherjee *et al.*, 2023). Self-efficacy affects hospitality abilities like communication, conflict resolution, and teamwork (Shah *et al.*, 2024). Competent complaint handlers are more likely to resolve consumer issues

constructively, building loyalty and reputation. Self-efficacy becomes more important as smart technologies like AI-driven customer service and data analytics grow more common. Employees must be confident in using these technologies to improve service. Self-efficacy increases job satisfaction and adoption of new technology, enhancing operational efficiency (Singh and Sihag, 2024).

Training and professional development are also needed since employee self-efficacy affects organizational outcomes. Comprehensive training improves employee abilities and self-efficacy, improving job performance (Joshi and Gupta, 2021). Hospitality organizations that prioritize employee development can improve service and build a resilient team in a competitive market. Self-efficacy greatly impacts hospitality employee performance and service quality. This section shows how self-beliefs affect workplace behavior using Bandura's theory (Bandura, 1978). The multimodal nature of self-efficacy might help hospitality firms conduct timely interventions to improve employee performance. Hospitality organizations can prepare their employees for industry changes by promoting self-efficacy (Wang *et al.*, 2024).

3.3 The Role of Smart Technology

As hospitality competitiveness increases, smart technology integration into service delivery becomes essential for efficiency and client experience (Mukherjee *et al.*, 2023). This chapter examines smart technologies in the hospitality sector, their effects on staff skills, and successful technology integration case studies (Youssef *et al.*, 2024).

3.3.1 Hospitality smart technology overview

AI, data analytics, and the Internet of Things (IoT) are among the smart technologies that the hospitality sector has quickly embraced (Mitra and Dey, 2024). By increasing consumer satisfaction and operational efficiency, these innovations provide distinctive value.

3.3.1.1 Machine learning, AI

AI encompasses computer programs that do human-like tasks like speech recognition, visual perception, language translation, and decision-making (Bhattacharya *et al.*, 2023). Data-driven system learning algorithms are developed by AI's machine learning subset (Ghosh and Jhamb, 2021). AI-driven chatbots offer round-the-clock client support, and predictive analytics maximizes hotel prices. AI-based chatbots can handle everyday customer service requests, thereby releasing the staff to handle complex issues (Singh and Sihag, 2024). Through faster response rates, this boosts customer satisfaction and productivity. Machine learning algorithms also use visitor information to produce recommendations that are relevant and improve visitor experience (Hossain *et al.*, 2023). AI is used in the hospitality industry in India, to ensure customer satisfaction and engagement (Nair and Senthil Kumar, 2024).

3.3.1.2 Business intelligence and data analytics

Data analytics utilizes systematic computer analysis to help make informed decisions. Hospitality businesses can identify trends, improve customer service, and streamline operations through analytics and business intelligence (Kar *et al.*, 2023). With consumer data, hotels can better market their products, control inventory, and manage visitors. Analytics technologies that analyze data from various sources to identify customer behavior and preferences enable personalized marketing campaigns and demand forecasts (Ghosh *et al.*, 2024b). This aspect enhances operational decisions since employees can be proactive in meeting client requests. Indian hospitality organizations are investing in advanced analytics solutions to make data-driven decisions (Joshi and Gupta, 2021).

3.3.1.3 IoT and smart devices

The Internet of Things (Mitra and Dey, 2024) allows devices to communicate and share data. IoT gadgets improve hospitality by making things easier and more personalized (Singh and Sihag, 2024). In-room electronic screens and smart

thermostats enable guests to customize their environment for comfort. IoT enables hotels to collect and evaluate visitor preferences for better customization. Monitoring room occupancy can automatically alter energy usage, enhancing sustainability and efficiency (Jena and Nayak, 2023). According to studies, such apps increase guest satisfaction in Indian hospitality. Smart appliances and inventory management systems reduce waste and increase efficiency in food and beverage businesses (Kumari *et al.*, 2024).

3.3.2 Improving employee capabilities

Smart technology in hospitality improve client experiences and employee skills (Manoharan *et al.*, 2024b). These tools automate mundane operations and provide real-time data to help employees perform more efficiently.

3.3.2.1 Streamlining procedures

Smart technologies streamline hotel and restaurant operations. Staff can focus on more meaningful service delivery by automating boring processes such as booking confirmation and payment processing (Kaur *et al.*, 2024). This efficiency increases staff contentment through inventiveness and customer centricity.

Property management system (PMS) software improves departmental communication, data access, and reservation handling (Shah *et al.*, 2024). Technology helps workers do jobs fast and accurately by decreasing burdens. Payment and inventory automation improve administrative efficiency, allowing staff to spend more time with visitors (Ghosh and Jhamb, 2021).

3.3.2.2 Enhancing decision-making

Smart technologies improve decision-making with actionable insights. Hospitality workers may make informed customer service, inventory, and operational decisions with comprehensive data analytics (Nair and Senthil Kumar, 2024). This encourages employee ownership and accountability.

Real-time data helps frontline staff serve clients quickly. Data analytics help managers uncover flaws and improve service quality and operational efficiency. Analytics help strategic planning by informing staffing, marketing, and resource allocation decisions, enhancing business performance (Ghosh *et al.*, 2024a).

3.3.3 Successful technology integration case studies

Real-world examples of technology integration in hospitality illustrate the transformative impact of smart technologies on operations and customer service.

3.3.3.1 *Example 1: AI in customer service*

A notable instance is the adoption of AI-based chatbots by major Indian hotel chains. One prominent hotel chain implemented an AI chatbot on its website to facilitate customer inquiries and bookings. This technology significantly improved response times and customer satisfaction, as evidenced by post-interaction surveys. Employees reported reduced workload, allowing them to focus on more personalized interactions. This case demonstrates how AI can enhance both operational efficiency and customer experience in hospitality.

3.3.3.2 *Example 2: Operational improvement with data analytics*

Another mid-sized luxury hotel in Mumbai used data analytics to improve operations. The hotel management studied occupancy statistics, seasonal trends, and visitor preferences using a business intelligence tool. This investigation optimized pricing models to boost off-peak occupancy (Seal and Gupta, 2024). The insights changed their marketing approach, resulting in demographic-targeted packages and a considerable rise in bookings during slow periods.

Finally, smart technologies are changing hospitality service delivery and management. Hospitality firms can improve staff skills, procedures, and guest experiences with AI, data analytics, and IoT. These technology advances will help companies meet changing market needs. The case studies show how technology integration improves operational efficiency and customer satisfaction in Indian hospitality (Youssef *et al.*, 2024).

3.4 Service Innovation

Hospitality service innovation is essential for competitive advantage and client satisfaction. Understanding service innovation is crucial for sustained growth as consumer expectations and technology change (Mehrotra *et al.*, 2024).

3.4.1 Hospitality service innovation definition

Service innovation involves creating or improving new or improved services, processes, procedures, and delivery methods to increase customer value and operational efficiency (Gupta *et al.*, 2025). This can include new services, advanced technology, or service process reengineering to improve client experiences in hospitality (Behera *et al.*, 2024). Service innovation affects personnel, management, and organizational culture as well as customer satisfaction (Singh and Sihag, 2024). Innovation in service models usually addresses consumer expectations for customization and digitalization. Service innovation helps hospitality companies stand out, attract new consumers, and retain old ones (Kar *et al.*, 2023). The digital transformation of India's hotel sector has elevated service innovation, creating new service delivery and consumer engagement models (Shah *et al.*, 2024).

3.4.2 Employee engagement in service innovation

Employee involvement is essential for hospitality service innovation (Al Halbusi *et al.*, 2023). Engaged personnel are more innovative and enthusiastic about innovation, improving service quality and customer satisfaction (Nair and Senthil Kumar, 2024). Employee engagement includes emotional, cognitive, and behavioral participation at work. Service innovation requires employees to feel valued and encouraged to suggest and implement new ideas

(Kaur *et al.*, 2024). Research shows that firms with high employee engagement achieve higher innovation results (Joshi and Gupta, 2021). An innovative service culture is developed by enthusiastic employees who own up to their assignments, attend training, and seek ways to improve (Behera *et al.*, 2024). Staff engagement is now regarded as one of the high priorities in promoting service innovation for the Indian hotel industry post COVID-19 pandemic (Mei *et al.*, 2024). Innovative training initiatives that encourage employees to change services are increasingly common (Ghosh and Jhamb, 2021). These approaches boost organizational and individual self-efficacy as well as service delivery.

3.4.3 Impact of innovative service models on self-efficacy

Service delivery by hospitality enterprises has changed owing to new innovative service models (Behera *et al.*, 2024). The impacts of these innovative service models regarding customization and innovation in service delivery have a crucial influence on both employee self-efficacy and service effectiveness (Dixit and Piramanaygam, 2023).

3.4.3.1 Personalizing guest experience

Personalized service refers to the customized offers to the needs and preferences of each visitor. Data analytics and artificial intelligence help hospitality businesses assess customer data and customize experiences (Singh and Sihag, 2024). Consumer data can help hotels improve client satisfaction and loyalty by customizing messages, activities, and changing room settings (Jena and Nayak, 2023). When employees get to apply visitor data for personalized service, they become more engaged and self-assured (Ghosh and Jhamb, 2021). Employees who see their work directly impact visitor experiences feel more competent and respected, boosting their confidence in providing exceptional service. This cycle encourages workers to try new things, creating an innovative culture (Hossain *et al.*, 2023).

3.4.3.2 Innovation in service delivery

Contactless services, smartphone check-ins, and virtual concierge services have emerged due to technological advances. COVID-19 expedited several of these advancements (Ghosh *et al.*, 2024b). These solutions boost operational efficiency and give workers cutting-edge tools to satisfy customers. Mobile apps make managing visitor requests easy, enabling faster responses and better service (Kumari *et al.*, 2024). As staff master these technologies, their confidence and self-efficacy rise, improving service performance. Encouragement to try new approaches creates a creative culture that boosts job satisfaction and self-efficacy by making people feel appreciated (Nair and Senthil Kumar, 2024).

3.4.4 Assessing service innovation's effect on self-efficacy

Service innovation's impact on self-efficacy depends on employee performance, guest satisfaction, and organizational success.

3.4.4.1 Assessments of competence

Competency assessments before and after new service models can identify employee self-efficacy changes. These assessments can include employee confidence surveys and performance evaluations in delivering new services or using new technologies (Shah *et al.*, 2024). In an Indian luxury hotel chain research, staff who received training on new service delivery practices showed enhanced confidence and job performance (Gupta *et al.*, 2025).

3.4.4.2 Staff surveys and feedback

Regular employee surveys on self-perception and engagement (Mei *et al.*, 2024) can reveal how service innovation affects employee self-efficacy. Employee perception surveys on innovation and job satisfaction can be useful (Jena and Nayak, 2023). Interviews and focus groups can also reveal employees service innovation experiences, helping organizations modify their support strategies.

3.4.4.3 *Guest satisfaction measures*

Service innovation's effects on self-efficacy must be assessed by measuring guest satisfaction. Guest satisfaction generally increases employee engagement and empowerment (Singh and Sihag, 2024). Organizations can link staff self-efficacy to guest experiences by monitoring consumer feedback, online reviews, and service quality measures. Hotels with innovative service methods and employee engagement had greater guest satisfaction (Al Halbusi *et al.*, 2023). This positive loop shows how employee self-efficacy affects service outcomes (Gupta *et al.*, 2025).

Service innovation is essential for hospitality competitiveness. Service innovation improves client experiences and operational efficiency by introducing new services, procedures, and delivery methods. Employee engagement is vital to this process since engaged employees innovate more. Innovative service models, especially individualized experiences and novel delivery methods, boost employee self-efficacy, enabling great service. Service innovation's impact on self-efficacy must be assessed using competency evaluations, employee feedback, and guest satisfaction indicators. Such tactics can help hospitality firms analyze their service innovation efforts and promote continual improvement to thrive in a changing sector.

3.5   Operational Efficiency and Self-Efficacy

Hospitality success depends on operational efficiency, which maximizes outputs while minimizing expenses and improving service quality. In this section the operational efficiency is discussed, its benefits for staff and ways to improve it, demonstrating its impact on hospitality self-efficacy.

3.5.1 Hospitality operational efficiency

Operational efficiency maximizes coordination, resource management, and service delivery in the hospitality industry in order to deliver great client experiences while reducing the consumption of resources (Gupta *et al.*, 2025). Operational efficiency is enhanced through process design, employee training, and technology. The COVID-19 pandemic highlighted the need for operational efficiency, forcing hospitality companies to adapt while preserving quality (Shah *et al.*, 2024). Technical integration is key to operational efficiency. Smart technologies such as PMS, POS, and mobile apps improve operations, communication, and resource allocation (Kumari *et al.*, 2024). Technology alone cannot ensure operational efficiency; it also requires a cohesive workforce that can execute operations efficiently and promote continual development. According to research, operational efficiency, customer satisfaction, and profitability are linked. Process improvements speed up service, reduce wait times, and improve client experiences, encouraging repeat business and positive word-of-mouth publicity (Kumari *et al.*, 2024). This dynamic is crucial for hospitality companies seeking to improve operational and employee performance.

3.5.2 Employee benefits of efficient operations

Hospitality workers' job satisfaction and self-efficacy benefit from efficient operations. Optimized processes help employees operate efficiently, lowering stress and improving working experiences (Nair and Senthil Kumar, 2024).

3.5.2.1 *Reduced workload and stress*

Operational efficiency simplifies roles and reduces workload. Streamlined check-in and check-out processes help front desk workers manage guest arrivals and departures, decreasing stress (Kumari *et al.*, 2024). Automating inventory management and scheduling frees up personnel to work on higher-value duties, making the workplace more enjoyable.

3.5.2.2 *Enhanced skill development*

Operational efficiency drives employee participation, therefore companies prioritize continual training and development (Ghosh and Jhamb,

2021). Skill upgrades improve technical skills and employee confidence, boosting self-efficacy. Good performers have higher job satisfaction and professionalism because self-empowerment boosts confidence in providing superior service.

3.5.2.3 Enhanced teamwork and collaboration

Efficient processes improve staff communication and collaboration. Process optimization improves staff collaboration and resource sharing (Kaur et al., 2024). Collaboration improves operational efficiency and builds employee confidence and self-efficacy. Strong cooperation improves workplace problem-solving and innovation (Jena and Nayak, 2023). Supported employees are more inclined to take initiative and solve problems creatively, enhancing their self-efficacy.

3.5.3 Improving operational efficiency strategies

Hospitality organizations have a wide range of strategies that they can use in order to achieve and sustain operational efficiency. Some of the most important strategies involve process optimization and staff empowerment, which are both of critical importance in enhancing the operational efficiency of an organization and, thereby, employee self-efficacy.

3.5.3.1 Process optimization

Process optimization improves service delivery by assessing and improving workflows. Lean management and Six Sigma methodology reduce waste and improve quality (Kumari et al., 2024). Process optimization in hotels can involve reassessing the guest journey from booking to check-out to find improvements. Automation in booking systems can improve client experiences and eliminate manual errors (Gupta et al., 2025). Optimized processes promote staff productivity and effectiveness, enhancing confidence and sense of achievement. Continuous feedback loops can help detect process bottlenecks and promote continuous improvement (Nair and Senthil Kumar, 2024). Feedback helps

employees understand their contributions and boosts self-efficacy.

3.5.3.2 Staff empowerment and autonomy

Employee autonomy improves operational efficiency, thus empowering them is crucial. Encourage staff to make decisions within their scope of work to boost job satisfaction and motivation (Kumari et al., 2024). Frontline staff who can resolve customer concerns or provide tailored services can better serve guests. Professional development helps people succeed in changing situations (Ghosh and Jhamb, 2021). Empowerment boosts self-efficacy, enabling employees to solve problems. Encouragement of idea development and employee participation in decision-making increases innovation since valued employees feel ownership and proactive in-service delivery. This empowerment boosts operational efficiency and self-efficacy.

Hospitality firms need operational efficiency to improve. In addition to improving service and visitor satisfaction, efficient operations foster employee growth. Reduced stress, improved abilities, and better teamwork boost employee self-efficacy. To improve operational efficiency, firms should optimize procedures and empower people. Hospitality organizations can increase service, employee satisfaction, and competitiveness in a dynamic market by refining systems and promoting employee initiative (Mehrotra et al., 2024).

3.6 Hospitality Sustainability

The global hotel sector prioritizes sustainability, which impacts environmental, social, operational, and employee engagement (Al Halbusi et al., 2023). The hospitality industry must embrace sustainable practices (Mehrotra et al., 2024) that support economic goals and social responsibility as climate change and resource depletion worsen. This section discusses the relevance of sustainability in the hospitality business, how sustainability programs affect employee self-efficacy, and how to create a sustainable workplace culture.

3.6.1 Sustainability's importance in industry

Hospitality sustainability includes environmental, social, and economic sustainability. Companies are realizing their responsibility in supporting sustainable practices to improve their image and attract environmentally concerned customers (Mehrotra *et al.*, 2024). Sustainable activities like energy conservation, trash reduction, and local sourcing can pay off. Energy-efficient solutions can lower utility bills, benefiting businesses and the environment. Consumer loyalty and market share are growing as customers prefer sustainable companies (Ghosh and Jhamb, 2021). Beyond economic benefits, sustainability promotes a responsible and ethical workplace culture. This transformation links hospitality organizations to the corporate social responsibility (CSR) movement (Nair and Senthil Kumar, 2024). Sustainability is not just ethical but essential for hospitality industry success.

3.6.2 Sustainability and employee self-efficacy relationship

Sustainability strategies and employee self-efficacy matter. An organization's sustainability efforts affect employees' self-efficacy, or confidence in their capacity to complete tasks and influence results. When employees see their companies proactively practicing sustainability (Mehrotra *et al.*, 2024), they take satisfaction in their work. For instance, waste reduction and energy conservation programs allow employees to meaningfully contribute to business goals (Kaur *et al.*, 2024). Research shows that environmentally responsible companies have happier, more committed workers. Staff are more engaged and creative when they believe their work improves society. Thus, sustainability boosts employee morale and self-efficacy while helping the environment and company.

3.6.3 Promoting a sustainable workplace culture

Training, community participation, and leadership commitment are needed to create a sustainable culture (Jauhari *et al.*, 2024). Following these measures can foster a sustainable workplace culture that boosts operational efficiency and employee self-efficacy.

3.6.3.1 Sustainability training

Creating a sustainable workplace culture requires comprehensive sustainability training. These programs should teach employees about sustainable practices, their benefits, and their roles in organizational sustainability (Ghosh and Jhamb, 2021). Diverse professions require personalized training to guarantee relevance and participation. Frontline workers may study energy saving, while kitchen workers could learn about food waste. Organizations empower employees to own sustainability efforts by giving them the skills and information.

In addition, implementing sustainability into new hire onboarding will instill these principles from the start. Staff participation in local clean-ups might strengthen their sustainability commitment and awareness (Gupta *et al.*, 2025).

3.6.3.2 Community awareness and engagement

Hospitality organizations can boost sustainability through community participation, including nonprofit partnerships, educational programs, and responsible tourism (Nair and Senthil Kumar, 2024). Local sustainability efforts like tree planting and farmer cooperation boost employees' job satisfaction and pride. This relationship enhances job satisfaction and self-efficacy by emphasizing their contributions. Community engagement establishes the company as a sustainability leader, attracting eco-conscious customers. Employees contribute to these activities, boosting the company's sustainability and enriching their own lives (Ghosh and Jhamb, 2021). Sustainability boosts brand reputation and operational efficiency in the hotel business. Sustainable methods empower a motivated workforce and improve the environment (Mehrotra *et al.*, 2024). Self-efficacy and job satisfaction increase as employees connect their work to sustainability goals, boosting productivity and innovation. To create a sustainable working culture, organizations should offer focused sustainability training and encourage community engagement (Mehrotra *et al.*, 2024). Engaging employees in meaningful sustainability efforts and providing appropriate knowledge will create a resilient, motivated staff dedicated to organizational and environmental goals.

3.7 Challenges in Enhancing Self-Efficacy

In the hospitality industry, the quality of services is directly linked to client satisfaction. Therefore, it is crucial to have a skilled, driven, and effective team. Employee worries, skill gaps, training limitations, and resistance to change are some of the obstacles that prevent employees from developing self-efficacy. These difficulties are covered in this section along with solutions.

3.7.1 Resistance to change

One of the main barriers to building self-efficacy in the hospitality industry is resistance to change. Frightened by new projects, technologies, or processes that management may introduce due to job insecurity or fear of the unknown, employees might resist new ideas (Gupta *et al.*, 2025). One of the common ways that this resistance shows up is through a disengagement from the training programs meant to improve skills and capabilities. The hotel industry is experiencing rapid and complex change due to shifting consumer preferences, market realities, and technology breakthroughs. Workers accustomed to conventional ways may find new methods as disruptive, leading to skepticism and a reluctance to embrace innovative practices. To minimize resistance, organizations must communicate the benefits of change clearly and involve staff in discussions of future projects. Involving employees in the change process will reduce their anxieties over the new processes and give them a sense of ownership (Ghosh and Jhamb, 2021). Leadership commitment also plays a critical role; leaders' enthusiasm over change sets an excellent climate for the staff (Öğretmenoğlu *et al.*, 2022).

3.7.2 Skill gaps and training constraints

One other significant barrier to building hospitality staff's self-efficacy is skill disparities.

Many workers might require new skills because technology changes so rapidly, and there is a different kind of expectation on customer care. Digital technologies keep increasing their adoption in hospitality management, it implies that staff have to be adept with the use of unfamiliar software (Nair and Senthil Kumar, 2024). These gaps in ability can be exacerbated by training limitations, such as time, money, or access to comprehensive programs. Employees who do not receive adequate training may feel less confident in performing their tasks, negatively impacting their self-efficacy. Research indicates that perceived lack of training opportunities diminishes employee engagement and commitment (Al Halbusi *et al.*, 2023), ultimately affecting service quality. Organizations should invest in targeted training and development programs aligned with current and future operational needs. Workshops, online courses, and mentorship initiatives can significantly enhance employees' skills and competencies. Fostering a culture of continuous learning encourages employees to pursue training opportunities, thereby improving self-efficacy.

3.7.3 Addressing employee concerns and fears

Employee concerns and fears can significantly impair self-efficacy. Common fears include performance reviews, job security, and meeting management expectations. These anxieties can lead to a fixed mindset, where employees believe they cannot improve or succeed in their roles. In high-pressure environments like hospitality, fear of negative feedback from customers or colleagues can exacerbate anxiety and diminish confidence. When employees feel undervalued or constantly scrutinized, their self-efficacy declines, resulting in lower job satisfaction and increased turnover intentions (Ghosh and Jhamb, 2021). Implementing supportive performance management systems that focus on constructive feedback and recognition of achievements can alleviate these concerns. Creating an open and safe environment for employees to express their fears without judgment is crucial. Regular check-ins and surveys can help gauge employee sentiments and proactively address potential issues (Nair and Senthil Kumar, 2024).

3.7.4 Strategies to overcome obstacles

To address these challenges, organizations can implement several strategies to enhance self-efficacy among employees in the hospitality sector.

1. *Open communication channels*: open communication with employees should be established to ensure that the employees feel comfortable voicing their concerns and providing feedback. Regular meetings, suggestion boxes, and anonymous surveys can facilitate open dialogue and foster a culture of trust (Gupta *et al.*, 2023).
2. *Change management programs*: develop formal change management programs that include employees in every step of the process. Clearly articulating reasons for changes, which would demonstrate their benefits, and employee involvement in decision-making can drastically reduce resistance and increase engagement (Gupta *et al.*, 2025).
3. *Customized training programs*: design customized training programs that are specific to the gaps in skills and also aligned with the goals of the organization. Providing a variety of learning methods, such as online modules, hands-on training, and peer mentoring, accommodates different learning styles and improves skill acquisition.
4. *Recognition and incentives*: develop recognition programs that appreciate the work of employees and their milestones. Recognition of efforts will increase morale and motivation, as employees believe in their capabilities.
5. *Supportive leadership*: the leaders should also practice self-efficacy by showing trust in the abilities of the employees. Providing mentorship, coaching, and other resources will increase the belief of employees in themselves and enhance professional development (Öğretmenoğlu *et al.*, 2022).

The improvement of the self-efficacy of the hospitality workforce is a critical tool in boosting engagement, productivity, and service quality. To encourage self-improvement, resistance to change, skill deficits, and employee concerns need to be addressed. Open communication, proper training, and effective leadership support can improve individual and organizational performance among employees.

3.8 Recommendations for Hospitality Organizations

Hospitality firms need to invest in technology, employee welfare, and professional development to compete. These strategies enhance service quality and staff engagement and self-efficacy to enhance organizational performance. Key hospitality organization recommendations are provided.

3.8.1 Developing robust training initiatives

Proper training programs are needed in building hospitality employees' proficiency, confidence, and other performance areas. Training programs that develop skill gaps should be prioritized and foster a culture of continuous learning in organizations.

3.8.1.1 Digital literacy programs

Digital literacy is required for hospitality workers to effectively use the technology. Comprehensive digital literacy courses would increase employee self-esteem and technology education (Nair and Senthil Kumar, 2024). Such courses should entail online reservation systems, CRM software, and social media marketing (Gupta *et al.*, 2023). Simulations and participatory exercises enable employees to practice with the use of technology in a non-threatening environment (Gupta *et al.*, 2025). The challenge requires training material to be updated periodically. Digital skills in employees can be improved by conducting regular assessment, and it will ensure that proficiency will be achieved in new apps and equipment.

3.8.1.2 *Continuous feedback mechanisms*

Implementing continuous feedback mechanisms is crucial for fostering growth and efficacy. Traditional annual performance reviews may not suffice in the fast-paced hospitality sector. Real-time feedback systems that facilitate open communication between supervisors and employees are more effective. Regular feedback enables employees to recognize their strengths and weaknesses, thereby enhancing their confidence and competency.

Mechanisms such as informal check-ins, peer reviews, and feedback tools promote ongoing discussions about performance and development. Training supervisors in delivering constructive feedback is also vital; effective feedback should be behavior-specific, action-oriented, and achievement-focused to encourage a culture of positive reinforcement.

3.8.2 Creating a culture of innovation

Establishing a culture of innovation is essential for hospitality organizations to remain competitive. Encouraging employees to contribute ideas fosters an inclusive environment where creativity thrives. Organizations should provide platforms for employees to propose improvements and hold brainstorming sessions or innovation challenges to stimulate creative thinking (Nair and Senthil Kumar, 2024). Leadership plays a critical role in cultivating an innovative culture by modeling experimentation and risk-taking (Öğretmenoğlu *et al.*, 2022). Professional development programs focused on creativity and problem-solving can empower employees to think critically and contribute to organizational growth (Gupta *et al.*, 2025).

3.8.3 Investment in technology and employee well-being

Investing in technology and employee well-being is vital for enhancing organizational performance. Upgrading technological infrastructure, such as implementing advanced property management systems and digital concierge services, can significantly improve operational efficiency and customer satisfaction (Ghosh and Jhamb, 2021). Moreover, prioritizing employee well-being through wellness programs – focusing on physical, mental, and emotional health – fosters a supportive work environment. Initiatives such as stress management workshops and flexible working arrangements enhance employee satisfaction and retention (Jena and Nayak, 2023). Promoting a healthy work–life balance further supports employee engagement and commitment, ultimately boosting self-efficacy. Regular assessments of employee well-being can identify areas for improvement, ensuring that initiatives meet evolving needs. By focusing on robust training programs, fostering an innovative culture, and investing in technology and employee welfare, hospitality organizations can enhance employee self-efficacy, leading to exceptional service quality.

3.9 Conclusion

The hospitality industry relies heavily on the competency and confidence of its workforce, making the enhancement of self-efficacy a critical success factor. Self-efficacy correlates with higher job satisfaction, reduced turnover, and improved customer service (Khilnani and Nair, 2022). Key challenges include resistance to change, skill gaps, and employee concerns. To address these issues, organizations should implement robust training initiatives, particularly in digital literacy, while fostering a culture of innovation and well-being. This approach empowers employees, enhances individual performance, and improves overall organizational effectiveness, leading to better customer experiences and a sustained competitive advantage.

The future of self-efficacy in hospitality is promising but requires continuous adaptation to emerging trends, particularly in automation and artificial intelligence. Organizations that prioritize employee well-being, including mental health and work–life balance, will enhance self-efficacy and engagement. A collaborative effort among stakeholders – employers, employees, industry associations, and educational institutions – is essential. Leaders should commit to employee development and well-being, while employees should embrace a growth mindset.

Industry associations must promote best practices, and educational institutions should align
curricula with industry needs (Nair and Senthil

Kumar, 2024). By working together, stakeholders can create a resilient hospitality industry
equipped to navigate future challenges.

References

Aggarwal, S., Singh, L.B. and Srivastava, S. (2024) Psychological empowerment an antecedent to career satisfaction: Modeling affective commitment as a mediator and resilience as a moderator. *Kybernetes*.

Al Halbusi, H., Al-Sulaiti, K., AlAbri, S., and Al-Sulaiti, I. (2023) Individual and psychological factors influencing hotel employee's work engagement: The contingent role of self-efficacy. *Cogent Business & Management* 10(3), 2254914.

Bandura, A. (1978) Self-efficacy: Toward a unifying theory of behavioral change. *Advances in Behaviour Research and Therapy* 1(4), 139–161.

Behera, B., Panda, R.K., Tiwari, B. and Chaubey, A. (2024) Understanding the drivers of innovative work behaviour among millennial employees in India's IT sector: Some exploratory research findings. *Journal of Asia Business Studies* 18(6), 1620–1646.

Bhattacharya, P., Mukhopadhyay, A., Saha, J., Samanta, B., Mondal, M. *et al.* (2023) Perception-satisfaction based quality assessment of tourism and hospitality services in the Himalayan region: An application of AHP-SERVQUAL approach on Sandakphu Trail, West Bengal, India. *International Journal of Geoheritage and Parks* 11(2), 259–275.

Dixit, S.K. and Piramanayagam, S. (eds) (2023) *Teaching Cases in Tourism, Hospitality and Events.* CABI, Wallingford, UK.

Farooq, R., Zhang, Z., Talwar, S. and Dhir, A. (2022) Do green human resource management and self-efficacy facilitate green creativity? A study of luxury hotels and resorts. *Journal of Sustainable Tourism* 30(4), 824–845.

Ghosh, P. and Jhamb, D. (2021) How is the influence of hotel internship service quality a measurable factor in student interns' behavioral intentions? Mediating role of interns' satisfaction. *Journal of Teaching in Travel & Tourism* 21(3), 290–311.

Ghosh, K., Sharma, D. and Malik, L.R. (2024b) Work calling and abusive supervision: Boon or bane in hospitality organization? *International Journal of Hospitality Management* 117, 103638.

Ghosh, P., Kaur, C. and Yu, L. (2024a) Faculty resilience and performance: The mediating roles of work engagement and affective commitment. *Journal of Hospitality & Tourism Education* 1–13.

Gupta, S., Priyanka, S. and Kumar, S. (2023) Evaluating e-leadership self-efficacy through social media efficacy and participation. *Management and Labour Studies* 48(4), 514–530.

Gupta, S., Kushwaha, P.S., Badhera, U. and Singh, R.K. (2025) Managing tourism and hospitality industry during pandemic: Analysis of challenges and strategies for survival. *Benchmarking: An International Journal* 32(4), 1360–1386.

Hossain, M.R., Bhatia, A. and Akhter, F. (2023) Rerouting tourism and hospitality in crisis: A systematic literature review and future research directions. In: Hassan, A., Sharma, A., Kennell, J. and Mohanty, P. (eds) *Tourism and Hospitality in Asia: Crisis, Resilience and Recovery*. Springer, Singapore, pp. 309–336.

Jauhari, H., Kumar, M. and Pandey, J. (2024) Impact of transformational leadership on service delivery behaviours of frontline service employees. *Journal of Retailing and Consumer Services* 79, 103816.

Jena, L. and Nayak, U. (2023) Organizational career development and retention of millennial employees: Role of job satisfaction, organizational engagement and employee empowerment. *International Journal of Organization Theory & Behavior* 26(1/2), 115–131.

Joshi, V.A. and Gupta, I. (2021) Assessing the impact of the COVID-19 pandemic on hospitality and tourism education in India and preparing for the new normal. *Worldwide Hospitality and Tourism Themes* 13(5), 622–635.

Kar, A.K., Choudhary, S.K. and Ilavarasan, P.V. (2023) How can we improve tourism service experiences: Insights from multi-stakeholders' interaction. *Decision* 50(1), 73–89.

Kaur, H., Sharma, S. and Guleria, S. (2024) Leveraging social media and technology in transforming travel for sustainable tourism: A bibliometric analysis. In: Talukder, M.B., Kumar, S. and Tyagi, P.K. (eds) *Hotel and Travel Management in the AI Era*. IGI Global, Hershey, PA, pp. 401–424.

Khilnani, L. and Nair, M. (2022) Employee turnover intentions and human resource practices in hospitality industry. *Journal of Management & Entrepreneurship* 16(1), 55–68.

Kumari, J., Singh, P., Mishra, A.K., Meena, B.P.S., Singh, A. *et al.* (2024) Challenges hindering women's involvement in the hospitality industry as entrepreneurs in the era of digital economy. In: Khang, A., Dutta, P.K., Gupta, S., Ayedee, N. and Chatterjee, S. (eds) *Revolutionizing the AI-Digital Landscape*. Routledge, New York, pp. 129–137.

Manoharan, G., Ashtikar, S.P. and Kumar, S. (2024a) Delineation of artificial intelligence in the hospitality and tourism industries. In: Talukder, M.B., Kumar, S. and Tyagi, P.K. (eds) *Impact of AI and Tech-Driven Solutions in Hospitality and Tourism*. IGI Global, Hershey, PA, pp. 20–42.

Manoharan, G., Rao, C.G., Ashtikar, S.P., Kumar, S. and Nivedha, M. (2024b) Voyage virtuoso: Artificial intelligence in transforming tourism. In: Kumar, S., Talukder, M.B. and Pego, A. (eds) *Utilizing Smart Technology and AI in Hybrid Tourism and Hospitality*. IGI Global, Hershey, PA, pp. 79–97.

Mehrotra, A., Agarwal, R., Awan, U., Walsh, S.T. and Yaqub, M.Z. (2024) Zero waste solutions in hospitality: Technology alignment and agile management practices for responsible consumption and production of food. *Journal of Sustainable Tourism* 1–31.

Mei, C.W., Konar, R. and Kumar, J. (2024) The role of AI chatbots in transforming guest engagement and marketing in hospitality. In: Nadda, V., Tyagi, P.K., Singh, A. and Singh, V. (eds) *Integrating AI-Driven Technologies Into Service Marketing*. IGI Global, Hershey, PA, pp. 595–620.

Mitra, A. and Dey, A. (2024) Redefining horizons: Examining the effects of the internet on India's tourism and hospitality sector. In: Nadda, V., Tyagi, P.K., Singh, A. and Singh, V. (eds) *Integrating AI-Driven Technologies Into Service Marketing*. IGI Global, Hershey, PA, pp. 537–558.

Mukherjee, S., Baral, M.M., Nagariya, R., Venkataiah, C., Rao, U.A. *et al.* (2023) Systematic literature review and future research directions for service robots in hospitality and tourism industries. *The Service Industries Journal* 43(15–16), 1083–1116.

Nair, A.L. and Senthil Kumar, S.A. (2024) Examining mediating and moderating influences among career competencies and leadership aspiration. *Journal of Management Development* 43(4), 571–590.

Nisar, Q.A., Haider, S., Ali, F., Gill, S.S. and Waqas, A. (2024) The role of green HRM on environmental performance of hotels: Mediating effect of green self-efficacy and employee green behaviors. *Journal of Quality Assurance in Hospitality & Tourism* 25(1), 85–118.

Öğretmenoğlu, M., Akova, O. and Göktepe, S. (2022) The mediating effects of green organizational citizenship on the relationship between green transformational leadership and green creativity: Evidence from hotels. *Journal of Hospitality and Tourism Insights* 5(4), 734–751.

Schunk, D.H. (2012) *Learning Theories: An Educational Perspective*. Pearson Education, Boston, MA.

Seal, P.P. and Gupta, P. (2024) Artificial intelligence in human resource management in hotels: A qualitative approach. In: Kumar, S., Talukder, M.B. and Pego, A. (eds) *Utilizing Smart Technology and AI in Hybrid Tourism and Hospitality*. IGI Global, Hershey, PA, pp. 277–290.

Shah, N., Bhatti, M.K., Saraih, U.N., Abdelwahed, N.A.A. and Soomro, B.A. (2024) The achievement of sustainable development and business success through rational management decision-making in a circular economy. *International Journal of Innovation Science* 16(5), 956–980.

Sharma, S.K. and Akram, U. (2024) Assessing subjective career success: The role of career commitment, career resilience and self-efficacy: Evidence from hospitality industry. *International Journal of Quality and Service Sciences* 16(1), 44–62.

Singh, R. and Sihag, P. (2024) Role of empowering leadership as a mediator between HPWPs and Gen Y employee engagement: Evidence from Indian hotels. *Journal of Hospitality and Tourism Insights* 7(4), 2287–2309.

Wang, Q., Azam, S., Murtza, M.H., Shaikh, J.M. and Rasheed, M.I. (2024) Social media addiction and employee sleep: Implications for performance and wellbeing in the hospitality industry. *Kybernetes* 53(12), 5972–5990.

Youssef, A.B., Dutta, P.K., Doshi, R. and Sajnani, M. (eds) (2024) *AI, Blockchain, and Metaverse in Hospitality and Tourism Industry 4.0: Case Studies and Analysis*. CRC Press, Boca Raton, FL.

4 Impact of Environmental Pollution on Food Security

Ayan Chatterjee[1]* and Ashi Ramavat[2]

[1]Medhavi Skills University, India; [2]SRM University Delhi-NCR, India

Abstract

Environmental pollution is crucial since it compromises with food safety globally by posing challenges to agricultural productivity, food accessibility, and quality. Rapid industrialization, fast urbanization, and unsustainable agricultural practices have led to soil, water, and air pollution. Heavy metals, pesticides, and microplastics predominantly constitute the pollutants that degrade the very fabric of arable land, thereby decreasing growth and yield of some important crops. Further, it might also be found in air, among others, to destroy the quality of arable land, and perhaps may reduce the general growth and yield of food crops. Airborne contaminants include particulate matter and greenhouse gases contributing majorly to damage our climate; they worsen the increasingly extreme weather events, which cannot keep crops alive given the changes in the usual trend of agricultural cycles and livestock production (yield reductions). Food chains' toxicity builds up as they consume the harmful substances, affecting both food safety and public health, and increasing the health risks for individuals and animals. Developing countries, even in general, are disproportionately affected by pollution – more so indigenous people – compounded by other factors like poverty, poor infrastructure, and restricted availability of healthy food.

Addressing such a big issue certainly necessitates a good deal of integration of approaches through which pollution control can be brought together with sustainable agricultural practices. Precision farming is one innovative avenue promising that things could be put right at the agricultural level through organic cultivation and an advanced waste management system. Policymakers, the academy, and many stakeholders should now step forward within common platforms and take steps to systematize the reaction to conclude effective strategies for enforcement, raising awareness, and enforcement of regulations and practices that place emphasis on environmental protection and sustainable food systems.

This chapter explores the complicated relationship between environmental pollution and food security; and explains various causes and impacts on food security while presenting potential solutions. The uniqueness of this chapter is informed by its scrutiny of problems caused by pollutants in agricultural ecosystems and food supply chains, thus paving the way for actionable sustainable food system development objectives for global food systems.

4.1 Introduction to Food Security and Environmental Pollution

The Food and Agriculture Organization (FAO) defined food security in 1996 as a situation where 'all people, at all times, have physical and economic access to sufficient, safe, and nutritious food to meet their dietary needs and food preferences for an active and healthy life'. The four critical dimensions of food security are availability, access, utilization, and stability. Availability implies food must be produced and continually available, access refers to the individuals' capacity to gain access to food, utilization emphasizes

---

*Corresponding author: ayan4189@yahoo.com

© CAB International 2025. *Municipal Waste Management: Policies and Strategies*
(eds S. Kumar *et al.*)
DOI: 10.1079/9781836990666.0004

the appropriate consumption and utilization of nutrients, and stability deals with the persistence of such conditions over time. The best of intentions toward attaining world food security notwithstanding, the issues of fast population growth, poverty, climate change, and conflicts are persisting bottlenecks to the advancement of most vulnerable regions (Godfray *et al.*, 2010). Environmental pollution is all-pervading and affects ecological systems, health, and productive agriculture. More broadly, it could be categorized under air, water, and soil pollution. These include industrial releases, vehicular exhaust, greenhouse gases, as well as several others that make up climate changes and respiratory disorders. Water pollution includes chemicals, runoff from agriculture, and raw waste flowing into rivers, lakes, and oceans. This degrades aquatic ecosystems and pollutes the sources of drinking water. The main causes of soil pollution include excessive application of fertilizers and pesticides and heavy metal accumulation in the soil, causing a decrease in soil fertility and crop yields. Such pollutants interfere with the natural system and last for decades, thereby worsening environmental degradation and food production problems.

It is through understanding the interrelation between environmental pollution and food security that one can approach the solution of these problems. Environmental pollution can directly affect agricultural productivity in the sense that it degrades soil quality, contaminate water resources, and changes climatic conditions that favor crop growth. For example, air pollution may lower photosynthesis in plants, whereas contaminated water and soil introduce harmful substances into the food chain, thereby threatening human health (Schipanski *et al.*, 2016). Conversely, unsustainable agriculture has been an environmental pollutant. Deforestation, excessive usage of agrochemicals, and the emission of greenhouse gases are examples. It points toward a need to combine efforts between the issues related to food security and pollution for proper mitigation policies (Tilman *et al.*, 2002). The threats facing human survival, and by extension, sustainable development must receive the outmost attention. With the world population estimated to reach 10 billion by 2050, the demand for food will increase even higher, thereby pressing natural resources

to uncommon levels. Sustainable solutions include the adoption of eco-friendly agricultural technologies, promotion of circular economies, and the implementation of pollution control measures (Pretty *et al.*, 2018). International cooperation and policy action are also required to reduce the impacts of pollution and ensure an equitable food future. These efforts align with United Nations Sustainable Development Goals in the areas of zero hunger, climate action, and environmental sustainability.

4.2 Types of Environmental Pollutants Affecting Food Security

Food systems are now getting affected in more ways due to so many pollutants such as those based mainly on the environment, those related to the health of human beings, and the disruption of agro-ecosystems and challenges in the agriculture. In light of all these, food quality was effectively reduced due to the many challenges of biotic environmental pollutants defined mainly as those having pesticide and herbicidal effects – this is more supportable across the world. Those that relate to poisoning – including in humans, generally from pesticides or herbicides – come in principle among food-related environmental pollution.

4.2.1 Pesticides and herbicides

4.2.1.1 Commonly used agrochemicals and their mechanism of action

In the realm of agriculture today, such pest control and weed control chemicals take on considerable significance because of the uses for which they were primarily developed. Essential components are herbicides and pesticides that are all employed to save agricultural crops from the threat of pests, insects, diseases, weeds, or unwanted plants. Probably the best known among them is glyphosate that systematically fights against the paternalism shikimate route, the essential metabolic route in plant growth, and atrazine, which hinders photosynthesis by preventing the electron transport chain within the chloroplasts

(Solomon *et al.*, 2001). These chemical compounds, specifically devised to have an effect on the specific biological processes, were not effective even though they hit the required target in the event of pests and weed resistance development. Such a twin condition creates a higher burden of accumulating chemicals on the agriculture ecosystem and leads to handling more of a concentration or going for other alternative compounds by the farmers. Thus, a continuous cycle of chemical is necessary. The resulting dependence on chemicals further calls for employing strategies of integrated pest management to minimize the importance of this aspect (Pimentel, 2005).

4.2.1.2 Ecotoxicological effects on nontarget organisms

Widespread and often indiscriminate usage of pesticides and herbicides has significant deleterious effects on nontarget organisms by upsetting the harmony of agricultural ecosystems. Bees and butterflies suffer even more as pollinators when subjected to intoxication by pesticides applications, which has led to endangerment in some populations, especially those that are most important for the pollination of numerous crops (Potts *et al.*, 2010). It leads to occasional fruit, vegetable, and nut crop yield declines that threaten food security and plant biodiversity with immediate impact. Also, aquatic ecosystems are in great danger since residues from agrochemicals are washed into rivers, lakes, and wetlands. Toxic to fish and other aquatic fauna, biocides also frustrate efforts to maintain fish production and other aquatic resources. Not to be forgotten are underground microbes essential to nutrient cycling, biodegradation, and maintaining soil architecture by loosening the soil matrix, compacted after intensive tillage (Wagg *et al.*, 2014). Therefore, it destroys the entire microbial community leading to loss in and reduction of soil fertility making it relatively more difficult to continue crop production year in and year out. Such catastrophic ecotoxicological impacts highlight the urgent need for eco-friendly alternatives and strict legislation to be enforced on the use of chemicals.

4.2.1.3 Residue accumulation in food crops and implications for human health

Agriculture residues, including pesticides, are often accumulated from food crops. They always end up having severe health risks to humans. Some of them, like glyphosate and atrazine residues, have been known to cause endocrine disruption, which alters hormonal systems, causing reproductive, developmental, and metabolic disorders (Mnif *et al.*, 2011). These residues have been linked to carcinogenicity, neurotoxicity, and other chronic disease reversals due to chronic dietary exposure. Groups that are particular vulnerable, such as children, pregnant women, and individuals with pre-existing health conditions, are most affected by these risks. Over and above people's direct dietary exposure, these hazardous materials of pesticides and herbicides may further accumulate through the food chains.

A holistic approach covers continual monitoring programs in different food products to maximize suitable levels of maximum residue and passing pesticide residue in food. Furthermore, integration of available and sustainable strategies ensures that risks such as crops that control pests, crop rotation, or the use of biopesticides decrease the reliance on organic or synthetic chemicals (Pretty and Bharucha, 2015). Instituting an educational factor is another advocacy output related to developing and advocating safer food consumption by sensitizing the public. Institutionalizing such practices rather than embodying the campaign from stakeholders is believed to ensure food security and sustainability in the environment.

4.2.2 Heavy metals

Heavy metals include elements that generally have relatively higher atomic mass and density. These metals usually present themselves by being toxic or non-biodegradable. Metals such as lead, mercury, and cadmium have gained much attention owing to their severe environmental contamination and bioavailability which potentially can impact humans and cause major ecological disasters.

4.2.2.1   Sources of contamination

Heavy metal contamination in ecosystems result from both natural and human activities, which are primarily contributed by the latter.

- *Industrial discharge*: industrial practice, including heavy metal smelting and producing chemicals, heavily discharge enormous amounts of metal into surrounding areas. Sometimes, these industrial effluents contain metals which include lead and cadmium while being mercury. These metals find a way through and settle in aquatic systems and ecosystems (Abubakar et al., 2024).
- *Mining activities*: mining processes, especially for metal ores, disturb the earth's crust and release naturally occurring heavy metals. Acid mine drainage is a widespread by-product of mining, which mobilizes metals into water systems and results in long-lasting contamination. Abandoned mines are a particularly significant source of heavy metal leaching into the surrounding ecosystems as they represent persistent sources of contamination (Hudson-Edwards, 2003).
- *Urban runoff*: heavy metals are a major contributor to the pollution of heavy metals in the urban areas through stormwater runoff. These sources include emissions, such as lead from older fuels and brake linings, construction debris, and waste disposed of inappropriately. Runoff carries these metals into water bodies where they settle in sediments and accumulate with time (Makepeace et al., 1995).

4.2.2.2   Specific heavy metals of concern

Lead (Pb): Lead is a widespread environmental containment that is primarily introduced into the environment through industrial process, lead-based paints, batteries, and vehicular emissions. Lead strongly binds to organic matter and clay particles in soils, but its bioavailability is highly dependent on soil pH and other conditions. Plants can absorb lead from their roots and transport it to shoots and leaves, although this is usually limited. However, in the crops grown on contaminated soils, lead accumulation within the edible parts poses serious threats to human health. Human exposure to chronic lead toxicity is associated with neurotoxicity, delays in the growth of children, and cardiovascular issues in adults through contaminated food and water (Storelli, 2008).

Mercury (Hg): Another reason for mercury pollution is coal combustion, mining, or simply burning of dust containing gold, as well as waste incineration. Spread underneath earth, this power upward to the surface, and when it is unlocked, it fills the air, water, and earth. The main concern is the change in inorganic mercury into methylmercury because of microbial activities in aquatic environments. Methylmercury is toxic and builds up in fish and other aquatic organisms; this would ultimately be transferred up the food chain. An increase in the levels of mercury in big fish, including tuna and swordfish, and increased mercury levels in animals high on the food chain jeopardizes wildlife and humans. The seriousness of this is that methylmercury is passed to humans through fish consumption and has neurological effects, mostly in the developing child (Bolan et al., 2024).

Cadmium (Cd): Major sources of cadmium pollution include phosphate fertilizers, industrial emissions, and wastewater irrigation. It may be a powerful phytotoxin affecting plant growth; photosynthesis and nutrient intake in plants are affected by it. Rice, being globally used, is much affected by the accumulation of cadmium since it is grown under water-logged conditions, which enhances the solubility of metals. Chronic exposure to ingestion of cadmium-contaminated rice and other foods causes diseases – namely, kidney vulnerability, demineralized bones, and increased cancer risks (Zhitkovich, 2011).

4.2.3   Microplastic

Microplastic can be defined as small plastic particles with a measurement of less than 5 millimeters. Generally, they come from the break-up of larger-sized plastic waste particle that occur during the process in the environment: physically, chemically, and biologically. Various sources give origin to these tiny pieces of plastics. They derive from industrial wastes, agricultural practice such as films for mulch, wastewater treatments, and many other consumer items, such as cosmetics, artificial fabrics, and personal care products (Geyer *et al.*, 2017). They are pervasive throughout terrestrial and aquatic ecosystems, posing some of the greatest challenges for environmental health.

4.2.3.1 *Effects on soil health*

Microplastic generally affect the health of soil by their alteration in physical, chemical, and biological properties. This affects soil structure and lowers its porosity and hampers the water-holding capacity of soil, which implies that there will be alterations in terms of the nutrient cycle and compromise soil fertility. In other words, besides exposure to toxic chemicals, microplastics may also serve as carriers of pathogens, thus further compromising quality in soil (Hodson *et al.*, 2017). It is important because it degrades the valuable diversity of microorganisms' interactions with soils, which are essential in maintaining the equilibrium of ecosystems by interacting with a range of soil microbial communities, hence weakens the microbial functional and diversity for agroeconomic resilience.

4.2.3.2 *Potential for human ingestion through the food chain and associated health risks*

Microplastic may enter the food chain through pathways such as their uptake by crops grown in contaminated soils, absorption by edible plants, and accumulation in aquatic organisms that are consumed by humans. These particles have been shown to penetrate plant tissue through root systems and transport mechanisms, especially in crops like carrots, lettuce, and wheat. This poses direct risks for human ingestion when these crops are consumed (Li *et al.*, 2020).

Microplastic are ingested by fish, shellfish, and zooplankton, among others, in aquatic systems. Once ingested, the particles bioaccumulate within individual organisms, leading to biomagnification up the food chain. This magnification can amplify the exposure of humans. Research has demonstrated that microplastics can cause physical damages to the lining of the gastrointestinal tract in humans, as well as provoking inflammatory responses (Wright and Kelly, 2017). Moreover, there are often persistent organic pollutants (POPs) within microplastics, including polychlorinated biphenyls (PCBs) and pesticide residues. There is a tendency to adsorb many of these toxins on the surface, making it most likely that any ingested chemical is toxic.

The long-term biological impacts of microplastics on human health could proliferate with physical and chemical risks. Several lab studies indicate that chronic exposure to microplastic can lead to oxidative stress, endocrine disruptions by invading the stress, and genetic damage. Oxidative stress is seen as the direct result of reactive oxygen species (ROS) and damages cell structures, mainly DNA. The pollutants come from the shedding of chemicals coming from plastic additives like bisphenol A (BPA) and phthalates that disrupt regulation in hormones and are associated with reproductive and development birth defects (Teuten *et al.*, 2009). In addition, a few microplastics with nanoscale dimensions and properties might be capable of passing via cell barriers to build up in organs and tissue, however, it is still unsure how this would affect life in future.

This further underscore the indispensable requirement for those novel and holistic strategies of coping with microplastics contamination. Thus, actions such as managing plastics better, advocating biodegradable alternatives, and advancing research on ecological and health impacts of microplastics are called for to keep the environment intact and secure the food supply. Furthermore, allowing the spread and impact of microplastics will significantly reduce this threat to human health and help build a bright and sustainable future.

4.2.4 Nutrient runoff

4.2.4.1 *Nitrogen (N) and phosphorus (P): sources from synthetic fertilizers and livestock waste*

Nutrient runoff, mostly driven through nitrogen (N) and phosphorus (P), is a large contributor to environmental pollutants, with extensive implications for meals protection. These vitamins predominantly originate from the overapplication of synthetic fertilizers in agricultural practices and from livestock waste that is improperly managed. Excess fertilizers applied to fields are regularly washed away by way of irrigation or rainfall, running into nearby water bodies. Similarly, concentrated animal feeding operations (CAFOs) produce significant quantities of nutrient-rich waste, which can leach into groundwater or runoff into surface water systems (Juncal *et al.*, 2023). These processes elevate the concentration of nitrogen and oxygen in aquatic ecosystem, leading to an imbalance in natural nutrients and a setup for ecological disturbances.

4.2.4.2 Eutrophication processes: algal blooms, hypoxia, and their impact on aquatic food sources

The influx of nitrogen and phosphorus into water bodies triggers eutrophication, a process characterized with the aid of excessive nutrient enrichment that promotes the rapid growth of algae. Algal blooms, specifically those resulting from harmful algal species, can produce toxins that contaminate water and pose dangers to aquatic lifestyles and human health. As these blooms decay, they consume large amounts of dissolved oxygen in the water, leading to hypoxic or anoxic conditions, normally referred to as 'dead zones' (Pretty *et al.*, 2001).

These oxygen-depleted zones critically affect aquatic ecosystems, reducing biodiversity and disrupting food chains. Fish, shellfish, and different economically important aquatic species are especially affected as they migrate to oxygen-rich regions and/or face population declines. The loss of those species at once influences food protection through decreasing the availability of fishery resources that many communities rely on for nutrition and livelihoods (Elser *et al.*, 2007). Addressing nutrient runoff calls for integrated control practices, including the optimized use of fertilizers, advanced waste management systems for livestock operations, and the implementation of buffer zones and wetlands to capture runoff earlier before it reaches water bodies. By mitigating nutrient runoff and its cascading outcomes, we can guard aquatic ecosystems, maintain biodiversity, and ensure the sustainability of food sources.

4.3 Impact of Air Pollution on Crop Production

4.3.1 Particulate matter (PM) and (O_3)

Air pollution is a significant hindrance for the production of crops since particulate matter (PM) and ozone (O_3) are two pollutants that are most important and responsible for changing plant physiological processes, which ultimately reduces agricultural productivity and endangers food security worldwide. The issue of air quality, however, has been rapidly worsening as a result of the increasing traffic emissions and now demands immediate attention in dealing with the problem (Van Dingenen *et al.*, 2009).

4.3.1.1 Physiological effects on plants: impairment of photosynthesis, stomatal conductance, and growth

Particulate matter, which includes great air-borne particles, can cover leaf surfaces and block sunlight, thereby impairing the photosynthesis process. Plant energy utilization in producing biomass is greatly reduced as a result of reduced incident sunlight due to PM particle deposition (Grantz *et al.*, 2003). Additionally, PM can harm the leaf structure and reduce chlorophyll content, further altering the photosynthesis. Over time, this could result in stunted growth, lesser biomass, and premature aging of plants.

Ozone (O_3), a secondary air pollutant, can get into leaf tissues by using stomatal penetration and may cause oxidative stress. Oxidative stress in turn includes disrupted stomatal conductivity, thereby lowering the efficiency with which water and nutrients are taken up. Exposure to O_3 may also lead to membrane injury and inactivation of enzymes as well as hinder plant growth, leading finally to lower crop productivity. This is possible since damage caused by ozone will produce reactive oxygen species (ROS), which in turn expedite cell decomposition and general plant health status (Feng *et al.*, 2015).

4.3.1.2   Regions with elevated PM levels and corresponding reductions in crop yields

Several studies highlight the unfavorable results of PM and O_3 on crop production. For example, in areas with high PM concentrations, such as the major parts of South Asia, crop yields of staple grains like wheat and maize had been proven to decline drastically. Research performed in India found that PM accumulation decreased wheat yields by about 20% in urban areas compared to rural areas with cleaner air (Pandya *et al.*, 2022). Similar tendencies were found in China, where PM pollution has negatively impacted the productiveness of rice paddies, a staple meal for millions.

Similarly, increased levels of ozone in agriculture regions of North America have resulted in visible decreases in maize yields due to reduced photosynthesis and enhanced cellular damage. Researchers in Europe have also documented substantial yield losses in soybean and potato crops following exposure to ozone. These studies highlight the pressing importance of controlling air pollution to protect global food security. The only hope to protect agriculture and ensure the sustainable production of food in a world where pollution is increasing lies in collaborative efforts, such as stricter regulations on emissions and adoption of cleaner technologies (Belis *et al.*, 2022).

4.3.2   Sulfur dioxide

Sulfur dioxide (SO_2) is another major atmospheric pollutant, and its negative effects on food security include a change in soil chemistry and damaging of sensitive crops. Its involvement in acid rain and the associated ecological impacts deserve serious attention.

4.3.2.1   Mechanisms of acid rain formation and its impact on soil chemistry

SO_2 emitted through industrial processes like burning fossil fuels reacts with water vapor and atmospheric oxygen to produce sulfuric acid. This acid precipitates in the form of acid rain, drastically lowering the soil pH levels and leaching critical nutrients such as calcium, magnesium, and potassium from the soil. Acidified

soils are deficient in these nutrients, which are essential for plant growth, and contain toxic elements like aluminum in higher quantities. These chemical imbalances affect root development and reduce resilience, thereby greatly affecting agricultural productivity (Driscoll *et al.*, 2001).

4.3.2.2   Effects on sensitive crops (e.g., legumes) and implications for biodiversity

Sulfur dioxide particularly harms crops, such as legumes, which are extremely sensitive to the changes of pH in soils. Impaired symbiotic relations between roots and nitrogen-fixing bacteria reduce nitrogen fixation in legumes, causing a diminishment of growth and yield in such crops. Crop productivity, thus, not only suffers from reduced yields but also causes distributed fertility in the soils for the other crops. Acid rain degradation of the soil also brings about a fall in biodiversity of the agricultural ecosystem. The sensitive plant species will not be able to survive in the altered soil conditions, thereby paving the way for invasive species and reducing the overall stability of the ecosystem. Loss of biodiversity affects food security because it limits the range of crops that can be sustainably cultivated and reduces the availability of ecosystem services essential for agriculture (Singh and Agrawal, 2008).

4.4   Water Pollution and Its Effect on Agriculture

Water pollution drastically undermines agricultural systems and directly threatens the world's food security. Pathogens not only are transferred to crops by contaminated irrigation water but also put the safety of crops at risk and threaten public health because the disease-causing microorganisms introduced into the food chain are ingested with crops. Over time, heavy metals in the sources of irrigation become deposited in crops, leaving residues that poison the food and human health. The problem is more complicated by emerging contaminants such as pharmaceutical and personal care products (PPCPs) pollutants in the environment, which continue degrading the soil and water quality even as they continue to disrupt basic soil microbial ecosystems that are a prerequisite for supporting plant growth. Their long-term effect on crop health and productivity therefore calls for prompt implementation of sustainability in water resource management to achieve food security as well as secure public health.

4.5 Soil Pollution and Its Consequences for Food Security

Soil pollution is the ultimate threat to global food security, which arises from the accumulation of contaminants like heavy metals, POPs, and excessive salts. Those degradation effects of pollutant on quality are through upsetting physical, chemical, and biological properties of this soil and its fertility, thus impending a capability to promote agricultural productivity and production. These risks include crop yield production, food contamination, and disruption in the supply chains that are primarily suffered by vulnerable people and regions largely dependent on subsistence agriculture. Additionally, penetration of pollutants in soil often leads to bioaccumulation and biomagnification that amplify health risks for humans and ecosystems (Chormare and Kumar, 2022). Eradication of soil pollution is thus crucial to safeguard food security and ensure sustainable agriculture systems.

4.5.1 Persistent organic pollutants (POPs)

POPs are chemical compounds that persist in the environment and build up in soil. These chemical compounds are generally introduced through pesticides, industrial effluent, and improper disposal of wastes. POPs disrupt the soil microbial ecology, which plays a vital role in enzymatic activity and nutrient cycling crucial for plant health.

For example, research indicates that exposure to POPs, such as polychlorinated biphenyls (PCBs) and dioxins, diminishes microbial diversity and nitrogen and phosphorus availability (Li et al., 2023). Crop uptake of the toxic substances present in contaminated soils leads to the entry of these substances into the food chain,

posing chronic health risks through endocrine disruption and carcinogenicity. Remediation of POP-contaminated soils includes advanced techniques like bioremediation, phytoremediation, and soil washing. These techniques, however, require much more resources and hence a more critical role of stringent regulations and pollution strategies (Lohmann et al., 2007).

4.5.2 Salinity and soil degradation

Soil salinity is one of the major causes of soil degradation and a significant threat to food security, often enhanced by unsustainable irrigation practices and climate change. High salt concentrations disrupt the osmotic balance, reducing the uptake of water and essential nutrients by plants. This leads to physiological drought, nutrient imbalances, and stunted growth. Saline soils are particularly damaging to sensitive crops like rice and wheat, which are staples for billions of people. In extreme cases, salinization leads to abandonment of agricultural lands, contributing to desertification. Economic losses from productivity declines due to salinity are significant, especially in arid and semi-arid regions where irrigation is essential (Cuevas et al., 2019).

In fact, integrated approaches should be considered when addressing soil salinity: utilizing salt-tolerant crop varieties, precision irrigation techniques, and the application of amendments such as gypsum to reclaim degraded lands while maintaining agricultural output.

4.5.3 Altered mechanism of *Rhizobium* bacteria

The role of *Rhizobium* bacteria is crucial for the biological fixation of nitrogen, a process important in replenishing soil nitrogen and planting leguminous crops. Soil pollution profoundly affects the efficacy of *Rhizobium* legume symbiosis. Heavy metals like cadmium, lead, and mercury interfere with the molecular mechanisms of *Rhizobium*, damaging the nodulation genes and weakening activity in the nitrogenase, the enzyme responsible for fixing nitrogen. Root exudates must also change to facilitate *Rhizobium* colonization, and these changes also affect the efficacy of nitrogen fixation. This intervention not only reduces the productivity of leguminous crops such as soybeans, lentils, and peas but also their contribution to sustainable agricultural systems (Wang *et al.*, 2024).

Restoration of the functionality of *Rhizobium* and improvement of soil fertility can be achieved by microbial inoculants, bioremediation techniques, and pollutant-specific soil treatments.

4.6 Climate Change as a Catalyst for Pollution

Climate change is an excellent catalyst of environmental pollution; such changes significantly raise the effects in food security with complex, interacting mechanisms. It has a rather important effect on agricultural and ecosystems' health as well as on human well-being when climate change and environmental pollution interacts.

4.6.1 Interactions between climate change and pollution

Most specifically, the interaction between climate change and pollution reflects the heightened incidents and magnitudes of extreme climatic conditions, such as heavy rains, hurricanes, and floods, which contribute to the enhanced mobilization and transportation of pollutants generated from agriculture, urban areas, and industrial setups into terrestrial and aquatic ecosystems. For instance, intense precipitation events result in the runoff of fertilizers, pesticides, and industrial chemicals into rivers and lakes, resulting in eutrophication, algal blooms, and hypoxic zones that threaten aquatic life and reduce water quality for irrigation and consumption (Kinney, 2018). Likewise, soil erosion during storms increases sedimentation with embedded contaminants degrading agricultural land and water bodies.

Meanwhile, extreme and prolonged droughts and heatwaves enhance the concentration of airborne pollutants like particulate matter and ground-level ozone that deteriorate air quality and cause direct damage to sensitive crops by oxidative stress. Increased global warming also enables stored pollutants like methane from melting glaciers into the environment, thus complicating existing sources of pollution. The synergistic interactions interfere with normal cycles, which intensify environmental stressors and weaken the very resilience of crucial ecosystems for producing sustainably (Ofremu *et al.*, 2024).

4.6.2 Vulnerability of food systems

The compounded stressors have increasingly shown vulnerability in food systems, disrupting crop resilience, soil health, and stability in food production. The polluted water and soil systems have adverse effects on plant growth by changing the availability of nutrient and introducing toxic substances that interfere with metabolic processes. For example, heavy metals like cadmium and lead from industrial runoff enter agricultural soils. When these are absorbed into the food chain, they pose huge risks to the health of plants and human safety (Hartmann *et al.*, 2015). Conversely, excessive nutrient loads from agricultural runoff create imbalances in the soil microbiome, reducing their capacity to support healthy crop growth.

Climate-related pollution increases the vulnerability of agricultural systems to biotic stressors, such as pests, pathogens, and invasive species which flourish in new climatic conditions. Besides reducing yields, such disruptions compromise the nutritional quality of food since stress conditions exchange the composition of key macronutrients and micronutrients in crops. The economic effects of these issues are significant enough that they not only raise food prices but also adversely impact the heavily marginalized and low-income populations, which is where these changes may hit very hard (Smith and Olesen, 2010).

To address this aggravated contribution, there has to be an integration of sustainable agriculture practices, pollution control measures, and climate adaption strategies. Policymakers, scientists, and local communities have to integrate and work effectively in order to mitigate such compounds and thicken probabilities toward ensuring stable and secure food supply.

4.7 Health Implications of Polluted Food Sources

Environmental pollution, therefore, significantly threatens food security through contamination of food sources and public health threats. Polluted agricultural lands become a reservoir of pathogens that are transmitted to the crops, hence causing food borne diseases. Contaminated produce becomes a vehicle for transmitting pathogens, resulting in outbreaks such as *Salmonella*, *E.coli*, and *Listeria*. The occurrences above raise essential public health issues related to crops grown in pollution environments, primarily in regions where food safety standards are weakly enforced (Koutsoumanis *et al.*, 2014). Apart from acute foodborne illness, heavy metals in degraded soils create a long-term health threat. Crop-growing on such soils accumulates toxic chemicals such as lead, mercury, and cadmium, which can remain in the human body for long. Chronic exposure has been associated with neurodevelopmental disorders for children, while adults suffering from renal damage, cardiovascular illness, and cancer have a higher prevalence of chronic conditions. Because toxicity from heavy metal is insidiously manifested, managing pollution in such agricultural ecosystems seems to be indispensable.

The health impacts of polluted food also have a huge cost in terms of the economy. Healthcare systems of low- and middle-income countries bear the highest burden in managing illnesses resulting from contaminated food sources. Furthermore, the loss in productivity due to absenteeism associated with diseases as well as lifelong health conditions means less economic activity. These factors all resonate together to give an impression that pollution of the environment, food insecurity, and socioeconomic stability have interlinking causes and so need to be collectively addressed (Rashid *et al.*, 2023).

4.8 Socioeconomic Factors and Vulnerability

Environmental pollution has the tendency to greatly enhance food insecurity among marginalized groups due to the increased exposure to pollutants and systemic inequalities. Vulnerable groups often stay in polluted regions where their health is compromised, further enhancing their vulnerability to food insecurity. Polluted water sources and poor soil quality decrease agricultural productivity, making it hard for these groups to access safe and nutritious food. In regions where industrial waste and pesticide runoff are rampant, food supplies may often become toxic, further endangering people's health who were already at risk of being poisoned (Bilali *et al.*, 2021). Socioeconomic disparities within food system further limit the availability of safe food and basic resources. Better-off populations are able to purchase better-quality food, seek quality healthcare to mitigate the risks of pollution-related health problems, or move to less polluted areas, while poorer groups are locked into environments with degraded food safety and availability. Such disparities perpetuate cycles of inequality, where a lack of clean resources directly impacts the well-being of vulnerable groups. More importantly, structural barriers such as lack of education, bad infrastructure, and limited political representation exacerbate these difficult challenges that leave marginal groups with fewer chances to change things in their favor.

Poverty plays a strategic role in increasing food insecurity as it restricts the ability of purchasing safer and nutritious food. Families living in poverty usually tend to eat cheaper and less nutritious food. Because food laws in most low-income areas of the world are not so strict, there is a change that the consumed foods are contaminated, thereby causing immediate unhealthy consequences as well as long-term developmental difficulties, creating intergenerational poverty cycles. In the agricultural sector, a primary source of livelihood in the rural areas, poverty and pollution combine to degenerate farmlands, limiting crop yields while further reducing the access to food (Pretty *et al.*, 2003). This then calls for some socioeconomic intervention factors to break these cycles and be able to find a balance within the equitable sustainable food systems by improving resource allocation, enforcing environmental law, and enhancing the capacity of vulnerable people to resist food insecurity and injustice.

4.9   Policy Responses and Mitigation Strategies

4.9.1   Regulatory framework for controlling specific pollutants

Environmental pollution, through a strong framework of regulation around the world, is mitigated with regards to its pervasiveness into food security. This can be reflected by one prominent example, the Stockholm Convention on POPs or the Global Treaty, eliminating or restricting production and use of specific hazardous chemicals. POPs are chemicals that have been known to be persistent in the environment, bioaccumulate up food chains, and cause harmful effects on human health and the environment.

Such a threat has serious implications for agricultural productivity and food safety. In this regard, the Stockholm Convention ensures that measures for phasing out these pollutants are strictly implemented so that food systems are not contaminated and agricultural is done safely. In addition to these efforts, regional initiatives complement them by dealing directly with localized problems and building compliance toward international standards, thus creating the multitiered defense against adverse impacts of pollution on food security (Weber *et al.*, 2011).

4.9.2   Sustainable agricultural practices

Sustainable agricultural practices should use some mitigative strategies such as the promotion and implementation of sustainable agriculture. These can help negate the impact on food security, especially due to environmental pollution. The adoption of integrated pest management and organic farming ensures minimal reliance on synthetic inputs while reducing soil and water contamination through the adoption of conservation agriculture. These methods focus on using natural resources efficiently, thereby eliminating chemical runoff and ensuring the ecological balance necessary for productive farming. In addition, agroecological practices through crop diversification, agroforestry, and enhancing soil health improve resilience against the effects of climate change and other environmental stressors besides mitigating pollutant impacts. Such practices ensure food production systems have a long existence while supporting attainment of global visions toward environmental protection and minimizing the pollution-induced dangers to food (Datta *et al.*, 2023).

4.10   Future Directions and Research Needs

Environment pollution and food security are connected issues that must be addressed within a robust framework. Interdisciplinary research, through scientist, policymakers, agriculturalists, and environmental experts, needs to come up with integrated solutions considering ecological, economic, and social dimensions. Efforts toward research should emphasize finding the source of pollution and the effects of this pollution on crop and livestock productivity, thus providing ways of mitigation. An example of this includes the application of biotechnology through pollution-tolerant crop varieties and advanced techniques for soil remediation. It can also facilitate the integration of socioeconomic studies to design policies that protect the most vulnerable communities affected by food insecurity. Finally, interdisciplinary collaboration can enhance the global monitoring system for pollution and food safety and foster more effective responses to emerging threats. In a world affected by pollution, interdisciplinary approaches are needed to bridge knowledge gaps and leverage diverse expertise (Mokari Yamchi *et al.*, 2019).

4.11   Conclusion

Environmental pollution poses a significant and multifaceted threat to global food security. This chapter highlights the extensive interplay between various pollutants – such as pesticides, heavy metals, microplastics, and nutrient runoff – and their detrimental effects on agricultural productivity, food safety, and human health. Air, water, and soil pollution not only impair crop yields and contaminate food supplies but also exacerbate socioeconomic disparities, disproportionately affecting vulnerable and marginalized communities. The impacts of pollution are further compounded by climate change, which intensifies pollutant runoff and destabilizes food systems. Emerging contaminants, such as pharmaceuticals and personal care products, present new challenges for food security that demand urgent attention.

Addressing these challenges requires a collaborative and interdisciplinary approach. Policymakers must implement and enforce stringent regulatory frameworks, such as international agreements on persistent pollutants, while promoting sustainable agricultural practices to mitigate pollution's effects. Researchers need to explore innovative solutions to reduce pollutant exposure and improve resilience in food systems. Communities and stakeholders must work together to advocate for equitable access to safe and nutritious food, particularly for those most at risk. Ultimately, ensuring food security in the face of environmental pollution is a shared responsibility. By prioritizing sustainable practices, investing in scientific research, and fostering global cooperation, we can protect the integrity of food systems and secure a healthier, more resilient future for all.

References

Abubakar, M.Y., Ahmad, K.B., Mathew, T.S., Shamsudden, R., Muhammad, H.M. *et al*. (2024) Heavy metal pollution in aquatic ecosystems: A review of toxic impacts and remediation strategies. *African Multidisciplinary Journal of Sciences and Artificial Intelligence* 1, 75–86.

Belis, C.A., Van Dingenen, R., Klimont, Z. and Dentener, F. (2022) Scenario analysis of PM2.5 and ozone impacts on health, crops and climate with TM5-FASST: A case study in the western Balkans. *Journal of Environmental Management* 319, 115738.

Bilali, H.E., Strassner, C. and Hassen, T.B. (2021) Sustainable agri-food system: Environment, economy, society, and policy. *Sustainability* 13(1), 6260.

Bolan, S., Padhye, L.P., Jasemizad, T., Govarthanan, M., Karmegam, N. *et al*. (2024) Impacts of climate change on the fate of contaminants through extreme weather events. *Science of The Total Environment* 909, 168388.

Chormare, R. and Kumar, M.A. (2022) Environmental health and risk assessment metrics with special mention to biotransfer, bioaccumulation and biomagnification of environmental pollutants. *Chemosphere* 302, 134836.

Cuevas, J., Daliakopoulos, I.N., de Moral, F., Hueso, J.J. and Tsanis, I.K. (2019) A review of soil-improving cropping systems for soil salinization. *Agronomy* 9(6), 295.

Datta, S., Hamim, I., Jaiswal, D.K. and Sungthong, R. (2023) Sustainable agriculture. *BMC Plant Biology* 23(1), 588.

Driscoll, C.T., Lawrence, G.B., Bulger, A.J., Butler, T.J., Cronan, C.S. *et al*. (2001) Acidic deposition in the northeastern United States: Sources and inputs, ecosystem effects, and management strategies. *BioScience* 51(3), 180–198.

Elser, J.J., Bracken, M.E.S., Cleland, E.E., Gruner, D.S., Harpole, W.S. *et al*. (2007) Global analysis of nitrogen and phosphorus limitation of primary producers in freshwater, marine and terrestrial ecosystems. *Ecology Letters* 10(12), 135–1142.

Feng, Z., Hu, E., Wang, X., Jiang, L. and Liu, X. (2015) Ground-level O_3 pollution and its impacts on food crops in China: A review. *Environmental Pollution* 199, 42–48.

Geyer, R., Jambeck, J.R. and Law, K.L. (2017) Production, use, and fate of all plastic ever made. *Science Advances* 3(7), e1700782.

Godfray, H.C.J., Beddington, J.R., Crute, I.R., Haddad, L., Lawrence, D. *et al*. (2010) Food security: The challenge of feeding 9 billion people. *Science* 327(5967), 812–818.

Grantz, D.A., Garner, J.H.B. and Johnson, D.W. (2003) Ecological effects of particulate matter. *Environment International* 29(2–3), 2013–2239.

Hartmann, M., Frey, B., Mayer, J., Mäder, P. and Widmer, F. (2015) Distinct soil microbial diversity under long-term organic and conventional farming. *The ISME Journal* 9, 117–1194.

Hodson, M.E., Duffus-Hodson, C.A., Clark, A., Prendergast-Miller, M.T. and Thorpe, K.L. (2017) Plastic bag derived-microplastics as a vector for metal exposure in terrestrial invertebrates. *Environmental Science and Technology* 51(8), 4714–4721.

Hudson-Edwards, K.A. (2003) Sources, geochemistry, and fate of heavy metal-bearing particles in mining-affected river systems. *Mineralogical Magazine* 67(2), 205–217.

Juncal, M.J., Masino, P., Bertone, E. and Stewart, R.A. (2023) Towards nutrient neutrality: A review of agriculture runoff mitigation strategies and the development of a decision-making framework. *Science of The Total Environment* 874, 162408.

Kinney, P.L. (2018) Interactions of climate change, air pollution, and human health. *Current Environmental Health Reports* 5(1), 179–186.

Koutsoumanis, K.P., Lianou, A. and Sofos, J.N. (2014) Food safety: Emerging pathogens. *Encyclopedia of Agriculture and Food System* 21, 250–272.

Li, L., Luo, Y., Li, R., Zhou, Q., Peijnenburg, W.J. *et al.* (2020) Effective uptake of submicrometre plastics by crop plants via a crack-entry mode. *Nature Sustainability* 3(11), 929–937.

Li, Y.-F., Hao, S., Ma, W.-L., Yang, P.-F., Li, W.-L. *et al.* (2023) Persistent organic pollutants in global surface soils: Distributions and fractionations. *Environmental Science and Ecotechnology* 18, 10031.

Lohmann, R., Breivik, K., Dachs, J. and Muir, D. (2007) Global fate of POPs: Current and future research directions. *Environmental Pollution* 150(1), 150–165.

Makepeace, D.K., Smith, D.W. and Stanley, S.J. (1995) Urban stormwater quality: Summary of containment data. *Critical Reviews in Environmental Science and Technology* 25(2), 93–139.

Mnif, W., Hassine, A.I.H., Bouaziz, A., Bartegi, A., Thomas, O. *et al.* (2011) Effect of endocrine disruptor pesticides: A review. *International Journal of Environmental Research and Public Health* 8(6), 2265–2303.

Mokari Yamchi, A., Alizadeh-Sani, M., Khezerlou, A., Zolfaghari Firouzsalari, N., Akbari, Z. *et al.* (2019) Resolving the food security problem with an interdisciplinary approach. *Journal of Nutrition Fasting and Health* 6(3), 132–138.

Ofremu, G.O., Raimi, B.Y., Yusuf, S.O., Dziwornu, B.A., Nnabuife, S.G. *et al.* (2024) Exploring the relationship between climate change, air pollutants and human health: Impacts, adaptation, and mitigation strategies. *Green Energy and Resources* 3(2), 100074.

Pandya, S., Gadekallu, T.R., Maddikunta, P.K. and Sharma, R. (2022) A study of the impacts of air pollution on the agriculture community and yield crops (Indian context). *Sustainability* 14(20), 13098.

Pimentel, D. (2005) Environmental and economic costs of the application of pesticides primarily in the United States. *Environment, Development and Sustainability* 7(2), 229–252.

Potts, S.G., Biesmeijer, J.C., Kremen, C., Neumann, P., Schweiger, O. *et al.* (2010) Global pollinator declines: Trends, impacts and drivers. *Trends in Ecology & Evolution* 25(6), 345–353.

Pretty, J. and Bharucha, Z.P. (2015) Integrated pest management for sustainable intensification of agriculture in Asia and Africa. *Insects* 6(1), 152–182.

Pretty, J., Brett, C., Gee, D., Hine, R., Mason, C. *et al.* (2001) Policy challenges and priorities for internalizing the externalities of modern agriculture. *Journal of Environment Planning and Management* 44(2), 263–283.

Pretty, J., Morison, J.I.L. and Hine, R.E. (2003) Reducing food poverty by increasing agricultural sustainability in developing countries. *Agriculture, Ecosystems & Environment* 95(1), 217–234.

Pretty, J., Benton, T.G., Bharucha, Z.P., Dicks, L.V., Flora, C.B. *et al.* (2018) Global assessment of agricultural system redesign for sustainable intensification. *Nature Sustainability* 1, 441–446.

Rashid, A., Schutte, B.J., Ulery, A., Deyholos, M.K., Sanogo, S. *et al.* (2023) Heavy metal contamination in agricultural soil: Environmental pollutants affecting crop health. *Agronomy* 13(6), 1521.

Schipanski, M.E., MacDonald, G.K., Rosenzweig, S., Chappell, M.J., Bennett, E.M. *et al.* (2016) Realizing resilient food systems. *BioScience* 66(7), 600–610.

Singh, P. and Agrawal, M. (2008) Acid rain and its ecological consequences. *Journal of Environmental Biology* 29(1), 15–24.

Smith, P. and Olesen, J.E. (2010) Synergies between the mitigation of, and adaptation to, climate in agriculture. *The Journal of Agriculture Science* 148(5), 543–552.

Solomon, K.R., Giddings, J.M. and Maund, S.J. (2001) Probabilistic risk assessment of cotton pyrethroids: I. Distributional analyses of laboratory aquatic toxicity data. *Environmental Toxicology and Chemistry* 20(3), 652–659.

Storelli, M.M. (2008) Potential human health risks from metals (Hg, Cd, and Pb) and polychlorinated biphenyls (PCBs) via seafood consumption: Estimation of target hazard quotients (THQs) and toxic equivalents (TEQs). *Food and Chemical Toxicology* 46(8), 2782–2788.

Teuten, E.L., Saquing, J.M., Knappe, D.R., Barlaz, M.A., Jonsson, S. *et al.* (2009) Transport and release of chemicals from plastic to the environment and to wildlife. *Philosophical Transactions of the Royal Society B: Biological Sciences* 364(1526), 2027–2045.

Tilman, D., Cassman, K.G., Matson, P.A., Naylor, R. and Polasky, S. (2002) Agricultural sustainability and intensive production practices. *Nature* 418(6898), 671–677.

Van Dingenen, R., Dentener, F.J., Raes, F., Krol, M.C., Emberson, L. *et al.* (2009) The global impact of ozone on agriculture crop yields under current and future air quality legislation. *Atmospheric Environment* 43(3), 604–618.

Wagg, C., Bender, S.F., Widmer, F. and Van der Heijden, M.G.A. (2014) Soil biodiversity and soil community composition determine ecosystem multifunctionality. *Proceedings of the National Academy of Sciences* 111(14), 5266–5270.

Wang, S., Wang, W., Rong, S., Liu, G., Li, Y. *et al.* (2024) Key factors and mechanisms of microplastics' effects on soil nitrogen transformation: A review. *Soil & Environmental Health* 2(4), 100101.

Weber, R., Watson, A., Forter, M. and Oliaei, F. (2011) Review article: Persistent organic pollutants and landfills – a review of past experiences and future challenges. *Waste Management & Research* 29(1), 107–121.

Wright, S.L. and Kelly, F.J. (2017) Plastic and human health: A micro issue? *Environmental Science and Technology* 51(12), 6634–6647.

Zhitkovich, A. (2011) Chromium in drinking water: Sources, metabolism, and cancer risks. *Chemical Research in Toxicology* 24(10), 1617–1629.

5 Assessing the Environmental Impact of Plastic Pollution in Tourism: A Bibliometric Analysis

S.G. Rama Rao[1], Chirra Baburao[2]*, Kolli Srinivasa Rao[1] and Sujayalakshmi Devarayasamudram[3]

[1]GITAM (Deemed to be University), India; [2]Gujarat National Law University (GNLU), India; [3]North Carolina Central University, USA

Abstract

Plastic pollution is considered a hazard to society, one of the most serious environmental problems in populous areas and regions where tourism is the main economy and where there is little infrastructure for domestic waste recycling. Thus, in this chapter, bibliometric analysis is considered a tool for analyzing the leading trends in research on the environmental aspects of plastic pollution-related tourism. This chapter provides a tangible overview of academic disciplines, responsible authors and their texts, and other research trends framed in the given dates. Existing academic production over the last two decades examined through Scopus data, allowing for the highlighting of the leading publication clusters on this topic and those whose works and authors have had the most impact. The data was analyzed using bibliometric tools Biblioshiny R and VOSviewer. The latter will help map global areas with the highest number of publications, originating publications, and most cited papers and recent trends in research concerning plastic waste (inherent and plasticizers, additives, and stabilizers, microplastics), circular economy/circularity concepts implemented and policy interventions. The results of the study show that intellectual research focuses on key areas such as the impact of plastic pollution on marine ecosystems; the challenges of single-use plastics in the circular economy; and the importance of multi-stakeholder interaction in sustainable waste management. Finally, the chapter discusses methodological issues of the reviewed works, such as quasi- and quantitative approaches, case studies, and policy analyses, etc. Therefore, this chapter does intend to summarize the existing knowledge so it can be used in research and policymaking to this end, so it serves as a guide to improve the implementation of sustainable touristic practices with lower impacts on the ecological footprint of plastic pollution as an environmental issue.

5.1 Introduction

Plastic pollution has become one of the most ubiquitous environmental crises of the 21st century, with plastic being detected in habitats from the deepest ocean trenches to the summit of Mount Everest (Napper *et al.*, 2020). The tourism sector, a key pillar of global economic growth, is both the victim and the engine of this crisis. Every year, millions flock to coastal hot spots, mountain getaways and urban centers, but this influx leads to plastic use that strains the limits of sustainability. Items such as bottles, packaging, and disposable personal protective equipment (PPE) are being used only once and creating large quantities of waste, often

---

*Corresponding author: bchirra@gnlu.ac.in

© CAB International 2025. *Municipal Waste Management: Policies and Strategies*
(eds S. Kumar *et al.*)
DOI: 10.1079/9781836990666.0005

ending up not in formal and structured waste management systems but in natural environments (Garcés-Ordóñez *et al.*, 2020; Aragaw *et al.*, 2022). For example, plastic litter driven by tourism found on beaches in Santa Marta, Colombia, constituted 80% of total debris, threatening marine biodiversity and local livelihoods (Garcés-Ordóñez *et al.*, 2020). Likewise, remote islands such as the Andaman and Nicobar Archipelago are experiencing ecological degradation due to tourism-related plastic waste, emphasizing the sector's footprint on a global scale (Krishnakumar *et al.*, 2020).

Despite its growing exposure, systemic relationships between tourism and plastic pollution remain under-researched. Most existing literature is limited to individual case studies or specific classes of pollutants related to microplastics in marine sediments (Reed *et al.*, 2018) or PPE waste during the COVID-19 pandemic (Aragaw *et al.*, 2022). Very few endeavors synthesize on a global scale, trends, regional divergences, or policy implications. To illustrate, marine debris species regularly report economic losses in coastal regions due to marine debris events (Jang *et al.*, 2014). Still, mountainous and freshwater ecosystems are critical habitats for nature-based tourism and are infrequently represented in these broader dialogues (Semernya *et al.*, 2017). Indeed, the contribution of circular economy principles to reducing tourism's plastic footprint remains barely studied despite evidence that waste reduction strategies can reconcile tourism development with sustainability objectives (Mofijur *et al.*, 2021).

5.1.1 Tourism's role in exacerbating plastic waste

Tourism contributes considerably to the larger plastic waste generation, especially in coastal and ecologically sensitive areas. The more tourists, the more they consume plastic that may contribute to littering, including single-use bottles, food packaging, and containers. For example, as Garcés-Ordóñez *et al.* (2020) report on the beaches of Santa Marta, Colombia, an important tourist destination, which were polluted with plastic debris and 80% of the littered material found was closely related to tourism

activities. Likewise, in remote island locations such as the Andaman and Nicobar Archipelago, tourism's plastic pollution could degrade ecosystems and put marine biodiversity at risk (Krishnakumar *et al.*, 2020). Tourism peaks add pressure to waste management as local infrastructure is often not designed for the extreme variations of plastic disposal and inappropriate disposal methods, and leakage to marine and coastal environments will occur (Nachite *et al.*, 2019).

The environmental and economic consequences of the plastic waste generated by tourism are significant. Marine debris events have led coastal tourism hotspots, like Geoje Island in South Korea, to report substantial revenues lost based on cleanup costs and declining visitors per capita connected to plastic pollution (Jang *et al.*, 2014). Additionally, the COVID-19 pandemic revealed the double-edged sword nature of tourism, as lockdowns led to a large reduction in plastic litter in regions like Lake Tana, Ethiopia, but increased single-use PPE waste, indicating the tourism sector's susceptibility to unsustainable practices (Aragaw *et al.*, 2022). The length of stay and a tourist's plastic consumption rate are positively correlated as shown in research by Qiang *et al.* (2020); therefore, targeted policies such as a ban on single-use plastics and community-led cleanups are vital to address tourism's plastic footprint. These results reinforce the need for approaches that insert circular economy indicators into tourism management, as they improve the balance between economic growth and environmental sanitation.

There is no single approach when studying plastic pollution in tourism; hence, the guiding research questions of this study are formulated to help us better understand the dynamics of plastic pollution from different standpoints. We first present the key trends in plastic pollution and tourist environmental research (RQ1), revealing how the field has evolved. Second, we want to identify the leading authors and cited works in this domain (RQ2), acknowledging the works that have defined the discourse. Third, we analyze the key themes in plastic waste management, circular economy, and tourist policy literature (RQ3), which provide the basis for strategies/frameworks to combat plastic pollution. Fourth, we explore the geographic distribution of academic studies on tourism-related plastic pollution (RQ4), investigating why certain regions have become hot spots for research. Finally, we provide a synthesis to evaluate how plastic pollution and sustainable tourism research might impact policymakers and industry actors in mitigating environmental impacts (RQ5), thereby bridging the gap between academia and practice.

This study aims to add to the emerging literature on sustainable tourism and climate conservation by answering these questions. Moreover, it seeks to offer practical recommendations for stakeholders to reshape the tourism development process to be in line with global sustainability ambitions, as the United Nations Sustainable Development Goals (SDGs), particularly the SDG 12 (Responsible Consumption and Production) and SDG 14 (Life Below Water). This study provides a bird's-eye view of the barriers and opportunities to tackle plastic pollution in tourism, thus enabling more meaningful and effective actions toward addressing plastic pollution in tourism issues through bibliometrics.

5.2   Literature Review

5.2.1   Plastic pollution in tourism and ecosystems

Plastic pollution is one of the most critical environmental problems today, and tourism is associated with the spatial expansion of plastic in different ecosystems. Once plastic enters the environment, much of it becomes virtually irreparable as it does not break down quickly, and there are few options for remediation, causing drastic and systemic ecological effects through changes in carbon and nutrient cycles, indoor habitat modifications, and harm to species diversity (MacLeod *et al.*, 2021). As tourism monopolizes this issue further, the influx of transient visitors brings exponentially more single-use plastics, which are often improperly disposed of and end up in oceans and on shores (Wabnitz and Nichols, 2010). Plastic waste enters the environment mainly through poor waste management systems, road runoff and wastewater pathways, which contribute to large quantities of plastics in aquatic habitats (Watt *et al.*, 2021). The ecological effects are

devastating, harming marine mammals, birds, and reptiles through entanglement and ingestion, indicating that more research is needed to quantify these effects and find ways to mitigate them (Quayle, 1992). Management approaches to date focus on mitigating plastic emissions by avoiding virgin plastic consumption and improving waste management through international collaboration (MacLeod *et al.*, 2021). These include value-added products using ocean plastics to form a global supply chain, and tackling the economic and ecological problems of plastic pollution (Watt *et al.*, 2021). However, an ecosystem-wide approach must be undertaken through economic incentive structures, legislation, and sustainable alternatives to combat plastic pollution and its effects on tourism and tourism-adjacent areas (Quayle, 1992).

5.2.2   Global distribution and sources of plastic pollution

Plastic debris is ubiquitous, from remote Antarctic sediments (Reed *et al.*, 2018) to high-altitude environments such as Mount Everest (Napper *et al.*, 2020). In coastal tourism hotspots, such as Santa Marta in Colombia, it is reported that 80% of litter found on beaches is from tourism (fixed and single, for example, bottles and packaging) activities (Garcés-Ordóñez *et al.*, 2020). Correspondingly, pristine islands in the Andaman and Nicobar Archipelago are polluted by tourism-associated plastics, endangering marine biodiversity (Krishnakumar *et al.*, 2020). In Slovenia, Laglbauer *et al.* (2014) recognized tourism as an important source of macrodebris and microplastics on beaches. In contrast, Retama *et al.* (2016) reported that microplastics were also found in Mexican tourist beaches, confirming the world's plastic pollution.

5.2.3   Environmental and ecological impacts

Microplastics and nanoplastics pose ominous threats in marine and terrestrial ecosystems. Mofijur *et al.* (2021), showed their ingestion by marine organisms, resulting in bioaccumulation and biomagnification, and Carlsson *et al.* (2021), detected microplastics in foraging habitats of Arctic walruses, suggesting more widespread contamination. Freshwater ecosystems are not immune;Laju *et al.* (2022), documented microplastics in lakes from India, and Shu *et al.* (2023), analyzed their migration through karst river systems in China. Lim *et al.* (2022), show that plastic pollution disrupts coral reefs near tourist zones.

5.2.3.1 Socioeconomic impacts and tourism

Plastic pollution erodes the foundations of tourism economies. Jang *et al.* (2014) estimated that Geoje Island, South Korea, lost $29–37 million in tourism revenue due to a marine debris incident. Qiang *et al.* (2020) showed that coastal pollution decreased tourist stay duration, which lowered local earnings. In contrast, the COVID-19 lockdowns led to transient reductions in plastic litter (Lake Tana, Ethiopia; Aragaw *et al.*, 2022) and enhanced reef conditions (Patterson Edward *et al.*, 2021), highlighting tourism's role in stimulating pollution.

5.2.3.2 COVID-19 and plastic pollution

Both mask and glove usage increased throughout the study. Aragaw *et al.* (2022) reported PPE debris on Ethiopian shorelines, whereas Ormaza-González *et al.* (2021) reported analogous trends for Ecuador. Some places experienced a temporary respite from debris buildup due to mass tourism during lockdowns, such as Bali, Indonesia, where sunk debris was reduced sharply (Suteja *et al.*, 2021). Both findings underscore the double-edged sword of tourism as a driver of pollution and a globalized phenomenon vulnerable to international disturbances.

5.2.3.3 Management and mitigation strategies

As in most areas of economics, strong waste management and strong policy are essential. Willis *et al.* (2018) showed that from strong policy waste abatement campaigns, 40% decreases in marine plastic influx were evident (in high policy areas). Citizen science projects such as Honorato-Zimmer *et al.* (2019) engage communities in litter monitoring, strengthening

stewardship. As proposed by Mofijur *et al.* (2021), they are focused on recycling and biodegradable options to reduce tourism's impact on plastic waste. For example, tourism development that balances environmental conservation is evident in the South African Grootbos Reserve, an example of localizing Sustainable Development Goals (SDGs) (Dube and Nhamo, 2021).

The literature demonstrates the far-reaching consequences of plastic pollution on ecosystems and economies, and tourism plays a major catalytic role. Gaps in pollution dynamics in understudied regions (e.g., mountain, freshwater ecosystems) remain despite regional studies offering valuable insights. Future research needs to prioritize interdisciplinary approaches, policy evaluation, and community-driven solutions to address the environmental legacy of tourism.

5.3 Research Questions

RQ1 What are the main trends in plastic pollution and tourist environmental research?

RQ2 What are the most prominent authors and cited works in this study area?

RQ3 What are the main themes in plastic waste management, circular economy, and tourist policy literature?

RQ4 What regions do the most academic study on tourism plastic pollution, and why?

RQ5 How can plastic pollution and sustainable tourism research help policymakers and industry reduce environmental impact?

5.4 Analysis

The data points from 1999 to 2025 show that this is a developing field (Table 5.1). The 2.7% annual growth rate, with 113 documents published across 45 sources, indicates significant ongoing interest from scientific communities motivated by rising global worries about plastic litter. A cross of emerging trends and historical context, the average document age is meaningful – it highlights studies of interest from the past 3.82 years, while foundational works (e.g., Gregory, 1999) also remain relevant. The average

Table 5.1. Descriptive statistics of the bibliometric dataset. Author's own

Description	Results
Timespan	1999–2025
Sources (journals, books, etc.)	45
Single-authored docs	2
References	7546
Keywords plus (ID)	1701
International coauthorships %	32.74
Documents	113
Document average age	3.82
Coauthors per doc	6.04
Average citations per doc	39.58
Authors of single-authored docs	2
Authors	655
Authors' keywords (DE)	412
Article	113
Annual growth rate %	2.7

number of citations per document was high (39.58), indicating the academic impact of the field with landmark studies like Garcés-Ordóñez *et al.* (2020) and Napper *et al.* (2020) shaping discourse. The analysis shows that collective action is in transcendence with an average of 6.04 coauthors per document and 32.74% international coauthorships, which align with the following initiatives to deal with transboundary-like challenges of marine microplastics (Carlsson *et al.*, 2021) and waste from tourism (Jang *et al.*, 2014). The large number of references (7546) and diversity of terminology (1701 keywords plus and 412 author's keywords) signal depth and interdisciplinary breadth with views across ecology, policy, and waste management. The virtually total absence of single-authored work (two documents) also reflects the collaborative spirit that characterizes contemporary environmental research. With this trend projected into the year 2025, the data shows a growing emphasis on finding solutions to microplastic pollution, models for the circular economy, and innovations in policy to develop sustainability strategies aligned with global goals. The dataset can bridge scientific validity with societal challenges; the practicability of planetary health is aligned with socioeconomic resilience.

The graph in Fig. 5.1 presents the annual number of published articles related to plastic pollution in tourism. As shown in Fig. 5.1, the research volume has increased substantially in the past few years, demonstrating the increasing academic interest in this environmental issue. At first, the annual number of articles was slight, which signified that researchers overlooked the intersection of plastic pollution and tourism. However, when the environmental impacts of plastic waste became apparent and the tourism industry's contribution to the problem was recognized, the scientific community became increasingly focused on this topic.

The upward trend in this graph indicates that researchers are increasingly recognizing the need to address plastic pollution in association with tourism. Several factors have contributed to this increase in scientific production, including increased public concern, stricter environmental laws, and global sustainability efforts. This literature is critical to help design sustainable strategies to reduce the environmental impact of plastic waste in tourist destinations. It further builds a foundation for decision-makers and industry stakeholders to take evidence-based action supporting sustainable tourism. In summary, the growing number of research papers underscores the need for ongoing studies and partnerships in tackling the multifaceted challenges of plastic waste in the tourism industry.

The graph in Fig. 5.2 presents the average citations per year covering the average number of citations per article annually regarding articles on plastic pollution in tourism. The data points express citation reversals because a piece of research has an uneven effect or relevance. Years with average citations as low as 1999 show that the topic at the academic level had not yet become mainstream. When awareness of environmental issues increased, especially near data years 2007–2015, there was a high citation average among environmental and interdisciplinary journals, with a peak in 2016. This peak presumably aligns with increased interest and urgency around tackling plastic pollution via global environmental campaigns and legislative conversations.

The next few years are a bit less clear in that it seems like there is a gradual decline around 2018, but that might suggest that they have

Annual Scientific Production

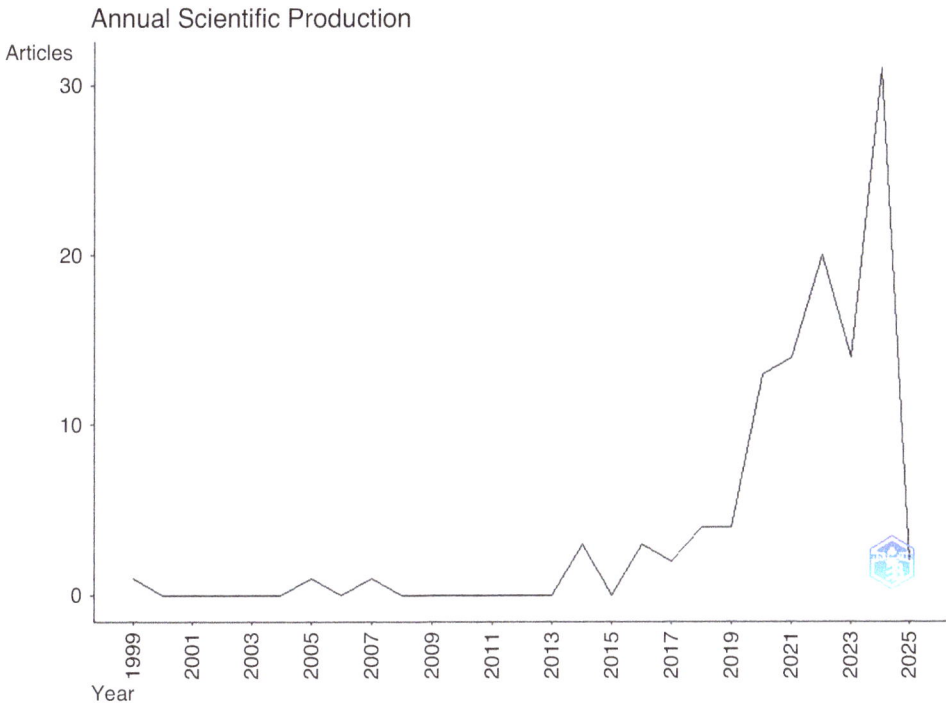

Fig. 5.1. Annual scientific production. Author's own.

explored initial hot topics and are shifting more toward new interests. But the trend looks to stabilize and rise once more into 2020 and beyond, signaling a new interest and knowledge of the significance of this research space. With upcoming research focusing on the subject, it is clear that plastic pollution in tourism will continue to be a widely discussed topic well into 2025. For example, citation trends highlight the dynamic nature of this research discipline and its critical contribution to informing sustainable practices and policies.

Figure 5.3 analyzes the common keywords found in academic references on plastic pollution and environmental monitoring, which can demonstrate important research trends and themes in the literature related to plastic pollution and environmental monitoring. The x-axis on this plot shows the number of occurrences of these keywords (the number ranges from 0 to 120), while the y-axis lists the most relevant keywords. The prominence of terms such as 'microplastic' and 'plastic' emphasizes their position as pollutants of concern, while

terms such as 'tourism' and 'marine pollution' indicate sector-based challenges. The much lower frequency of 'environmental monitoring' suggests a need to expand methodological frameworks to fill identified gaps in tracking and mitigating pollution. This distribution aligns with international research trends where microplastics and tourism-generated waste take priority because of their immediate risk to ecosystems and human economies. The exposed principle of circular economy approaches, and interventions in policy reveals the potential most vulnerable areas for reducing plastic leakage. Moreover, future studies may cover areas less studied, such as mountain or inland freshwater systems, to diversify environmental monitoring and management.

Figure 5.4 showcases the academic institutions and their countries, reflecting the widespread international publications on plastic pollution in tourism. Some key contributors from associated institutions include the Universidad Católica del Norte in Chile, Boston College (USA), and the University of Florence (Italy).

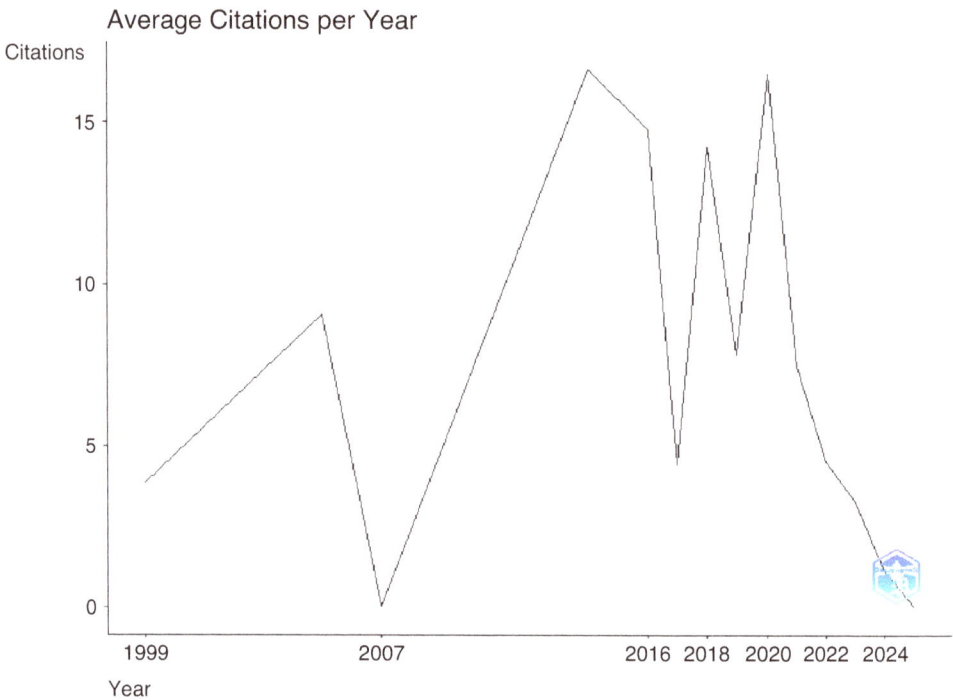

Fig. 5.2. Average citations per year. Author's own.

This serves as a reminder that the environmental implications of plastic pollution in tourism are not limited to one geographic area, shaping responses from various countries. China, India, Indonesia, and Brazil make up most of it, but that probably says more about their huge environmental challenges (due to high tourism activity and plastic waste generation) than the countries themselves. The inclusion of the US, Italy, and UK institutions also signifies the involvement of developed nations in being active players in research, contributing to high-end studies and policies. The participation of institutions in Ecuador, Mexico, and Sri Lanka, among others, indicates that even smaller or developing nations recognize the relevance of this issue and add to the worldwide knowledge pool. The geographic diversity is critical in creating more comprehensive and context-specific solutions to plastic pollution in this sector, as it introduces different perspectives and localized knowledge to the research. As this data shows, this is a global problem which needs a coordinated response to

minimize the environmental impact of plastic pollution in the tourism sector.

Figure 5.5 shows a bubble chart displaying the most relevant sources in respect to the number of documents published in a specific research domain. The number of documents is given in the *x*-axis, and the academic sources influencing the field is given in the *y*-axis. Dots are scaled according to the number of publications from each source; larger dots correspond to more relevant documents. The figure indicates that *Innovations In Pollution Research* is the most influential source, with 30 publications, which is much more than others. After this, the *Journal Of Environmental Management* ranks second, with 17 documents, which has a, very important role in the articles related to the environment. Another big contributor is *Environmental Pollution* with seven documents. With only four publications each, *Environmental Sciences and Pollution Research* and *Biological Control Information* are also moderately relevant to this research area. The rest of the sources

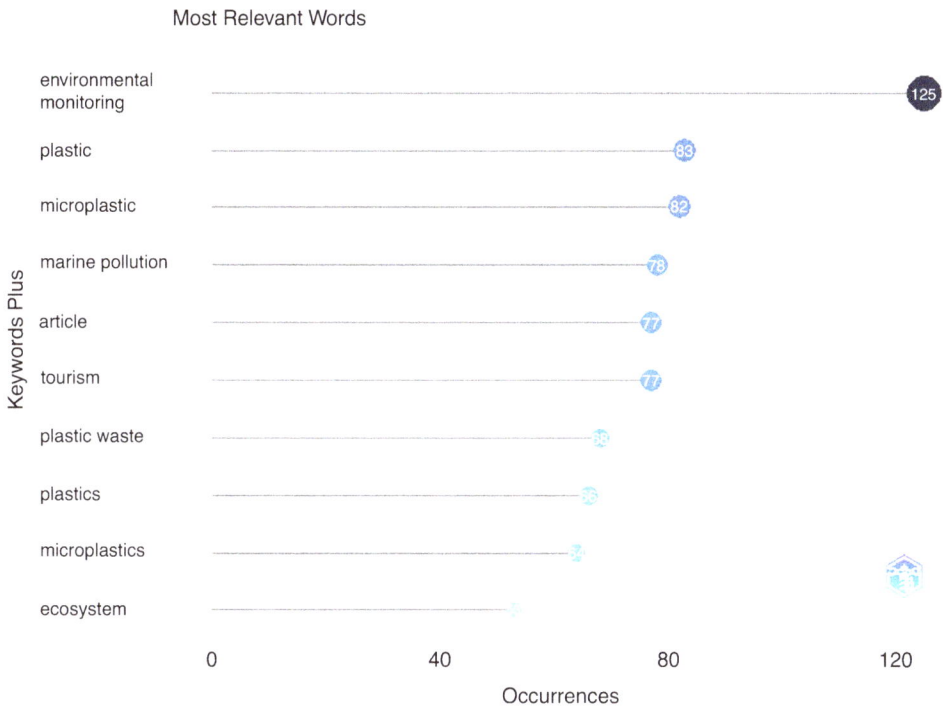

Fig. 5.3. Most relevant words. Author's own.

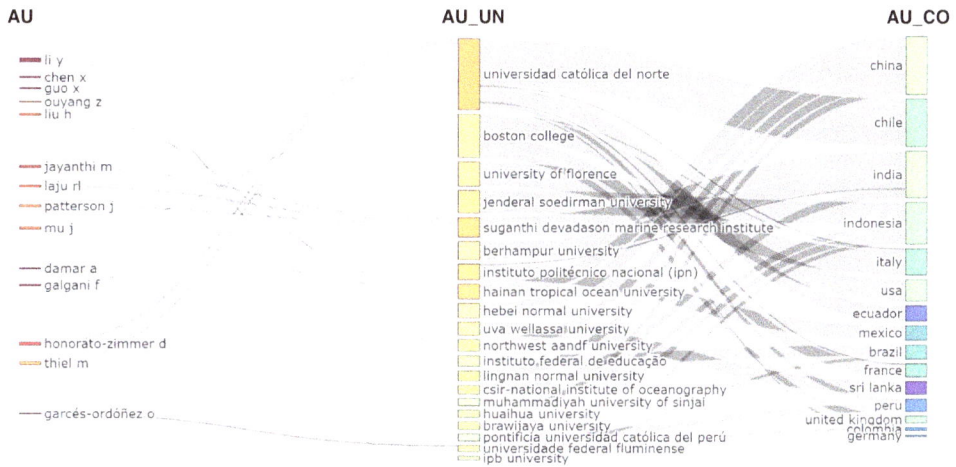

Fig. 5.4. Three-fold plot. Author's own.

Most Relevant Sources

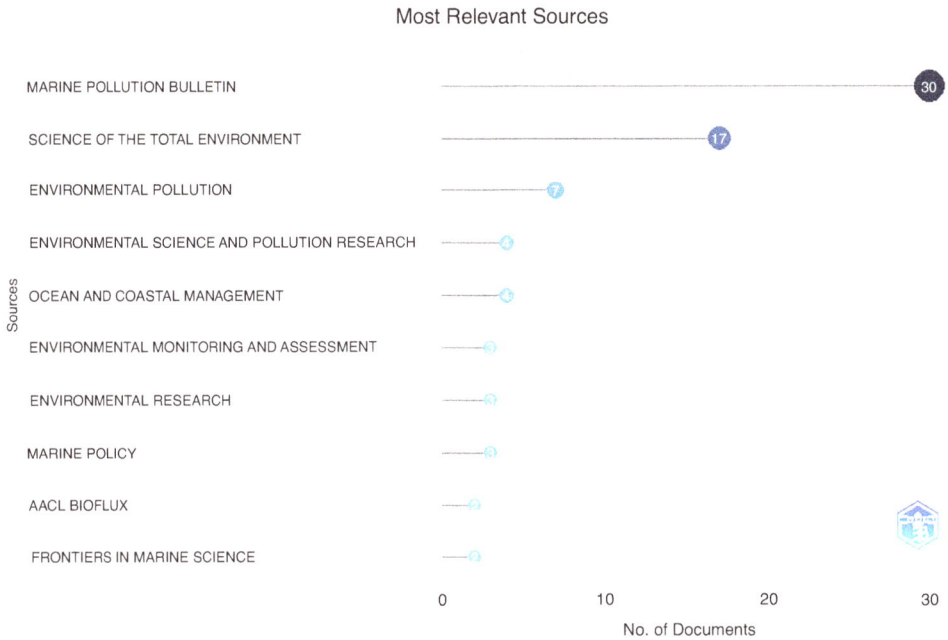

Fig. 5.5. Most relevant sources. Author's own.

present lower acquisitions with few documents counts from one to three, such as *Environmental Engineering Research*, *Environmental Research*, *Marine Policy*, *Aqua Hustler*, and *Progress in Marine Science*.

The distribution pattern clearly shows that a very small number of journals dominate the publication landscape, while most others contribute in smaller volumes. Indicating the need for more research published in key environmental and pollution-related journals in order to gain the most impact of the knowledge created. These top journals should be the priority for researchers who hope to publish or look for relevant literature in this field to reach as many academics as possible.

The application of bibliometric analysis illustrates Bradford's Law via the core sources. Bradford's Law is applied to identify the most productive sources (journals, conferences, etc.) by categorizing the sources in various zones that decrease productivity in a particular research field (Fig. 5.6). The x-axis is the logarithm of the sources' rank (Source log(Rank)), and the y-axis is the articles posted by these sources. The raindrop curve starts with a steep slope, corresponding to a few very high half-life core sources contributing significant parts of the literature. The curve flattens and indicates additional sources with fewer articles when looking at higher ranks.

The 'Core' zone of the graph illustrates the most influential sources that publish most of the articles on the topic. As there are core sources of data scientist studies, in the most relevant and widely cited studies, the following zones correspond to sources that contribute incrementally less. This analysis allows researchers to pinpoint the journals or conferences that dominate the research area, including plastic pollution. Simplifying the number of studies to follow helps readers keep track of new inventions, ensuring their work aligns with the best research in the field. Such a collaborative approach also helps us appreciate knowledge distribution and the influence of different kinds of publications in the research community.

Table 5.2 presents the h-index, g-index, m-index, total number of citations (TC), number of papers published (NP), from year of publication (PY_start), etc. *Marine Pollution Bulletin* is the most influential source, which

Core Sources by Bradford's Law

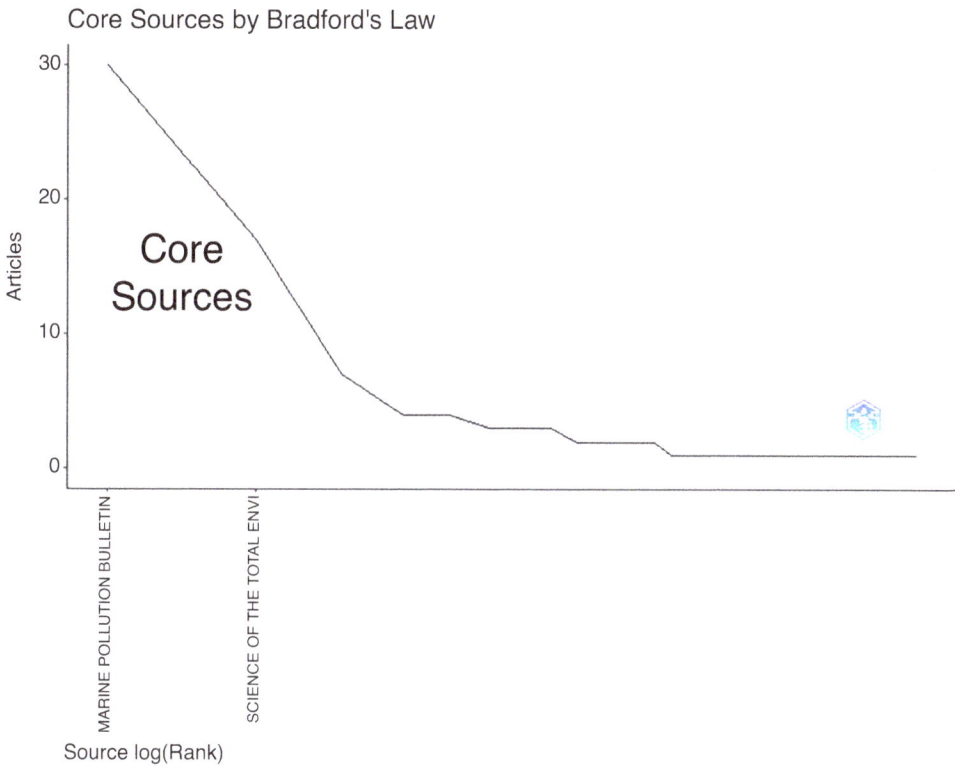

Fig. 5.6. Bradford's Law. Author's own.

Table 5.2. Sources local impact by h-index. Author's own.

Source	h_index	g_index	m_index	TC	NP	PY_start
Marine Pollution Bulletin	14	30	1.17	1557	30	2014
Science of the Total Environment	13	17	2.17	463	17	2020
Environmental Pollution	4	7	0.44	184	7	2017
Ocean and Coastal Management	4	4	0.15	451	4	1999
Environmental Science and Pollution Research	3	4	0.6	43	4	2021
Environmental Monitoring and Assessment	2	3	0.25	69	3	2018
Environmental Research	2	3	0.33	126	3	2020
Frontiers in Marine Science	2	2	0.4	49	2	2021
Marine Policy	2	3	0.25	150	3	2018
AACL Bioflux	1	1	0.17	2	2	2020

computed an h-index of 14 and a g-index of 30 with 1557 citations observed since 2014. An alternative option is *Science of the Total Environment*, which has a high m-index of 2.17 and 463 citations since 2020. *Environmental Pollution* and *Ocean and Coastal Management* show moderate impact; the latter's low m-index indicates less influence each year despite a long publication history since 1999.

Most Relevant Authors

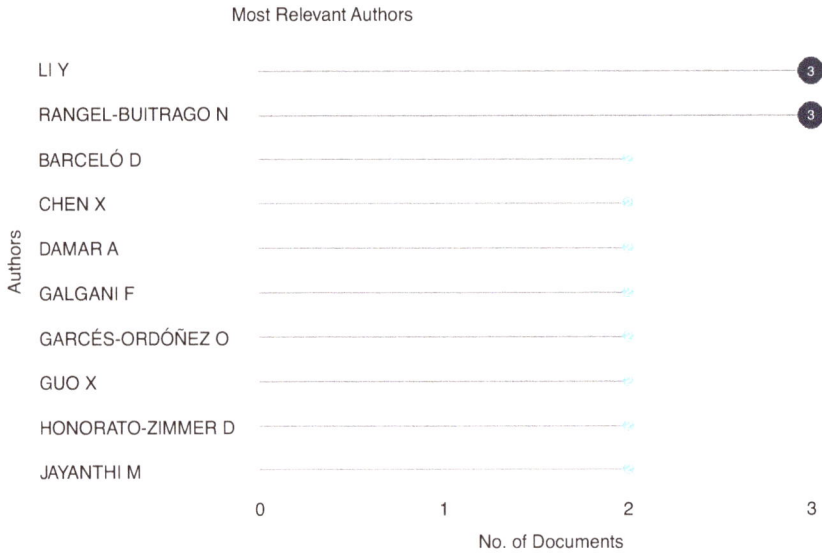

Fig. 5.7. Most relevant authors. Author's own.

The emerging potential is indicated by other sources, including *Environmental Science and Pollution Research*, *Frontiers in Marine Science*, and *Environmental Research*, which had relatively higher m-index values, suggesting recent activity and growth in bibliometric analysis. Such impact metrics are low for sources such as *Environmental Monitoring and Assessment*, *Marine Policy*, and *AACL Bioflux* because they have been minimally cited or have limited influence. *Marine Pollution Bulletin* and *Science of the Total Environment* are overwhelmingly the most influential overall, and other sources differ in their impact, with variability that some show potential for growth in the future. It enables researchers to track what they need to read and where to place their work.

The bibliometric analysis of major researchers in this study domain, presumably plastic pollution or environmental research, is given in Fig. 5.7. Authors are represented on the *y*-axis, while the total number of published documents is on the *x*-axis. The volume of documents per author is visualized by the size of each circle, where a higher volume corresponds to a larger circle.

Most authors contributed around two documents, with a few excelling with three publications. This indicates a relatively balanced spread of research contributions in this area, with no one author completely overwhelming the literature. Since many authors with similar publication counts, this is also a sign of collaborative and distributed research effort. This bibliometric information is disadvantageous for exploring niche contributions and promising research collaborations. It also furthers author contributions, enabling scholars to identify prominent researchers and significant studies in the area.

Figure 5.8 shows the development of plastic pollution research from 1999 to 2025 as, broken into two distinct phases. The studies from 1999 to 2022 primarily explored overarching categories (i.e., 'plastics' similar to Gregory, 1999, showcasing abundances of plastic debris along South Pacific beaches in the late 1990s) covering mass production and usage or high-level environmental impacts. During this time, the subcategory 'plastic waste' gained traction, indicative of an increasing focus on waste management issues, as represented by Santos *et al.* (2005), who connected tourism behavior to generating litter on beaches. From 2023 to 2025, research became highly specialized, where a search for 'plastic waste' further shortened to 'microplastic', underscoring the most urgent nature of smaller plastic particles'

1999–2022   2023–2025

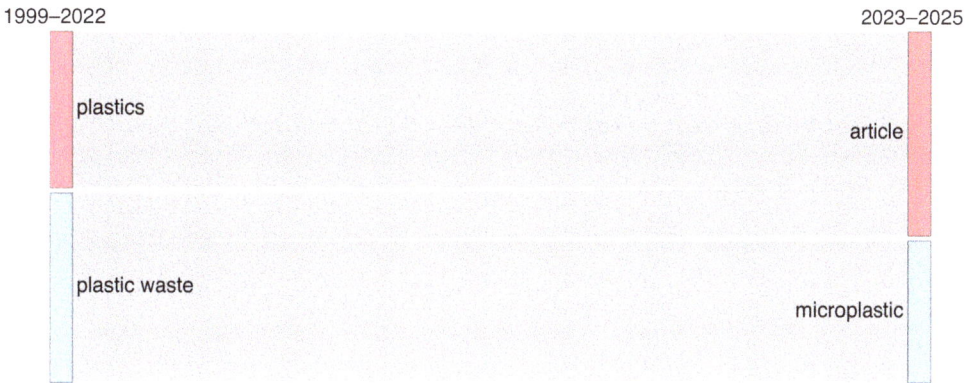

Fig. 5.8. Evolution of studies on plastic pollution between 1999 and 2025. Author's own.

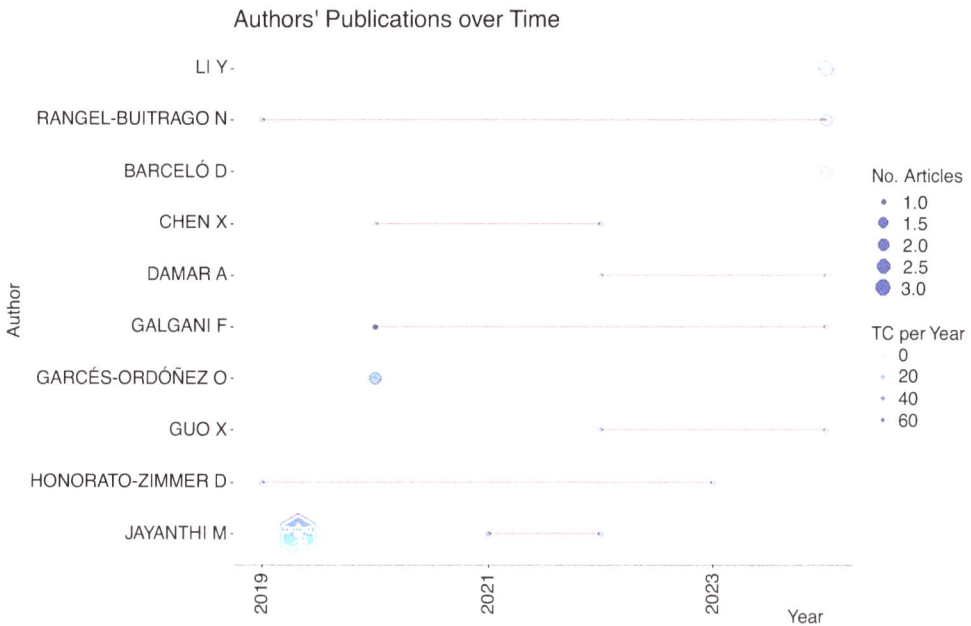

Fig. 5.9. Authors' publications over time. Author's own.

ecological and health effects, as in studies by Carlsson *et al.* (2021) regarding Arctic eco-systems and Shu *et al.* (2023) in rivers in karst landscapes. Simultaneously, the persistence of 'article' under 'plastics' reflects interest in both circular economy pathways (e.g., Mofijur *et al.*, 2021) and policy-oriented mitigation strategies (e.g., Willis *et al.*, 2018). This timeline illustrates a paradigm transition from macroscale plastic research to microplastic-focused studies,

propelled by technical progress and international sustainability strategies. The data available until October 2023 highlight that there will be a focus on people, pro-planet economies and food systems to curb plastic leakage, eco-design sustainable materials and products, and sustainable companies worldwide, especially following the UN Sustainable Development Goals SDGs.

Figure 5.9 shows the number of publications (i.e., scholarly output) and their citation impact over the years by the leading authors in research about plastic pollution. Some of these authors are Rangel-Buitrago and Galgani, with numerous articles and elevated TC per year, which denotes these authors as key authors in marine pollution. Notably, some authors, such as Garcés-Ordóñez and Jayanthi, exhibit clear activity in recent years, consistent with the increasingly important subject of tourism-driven plastic waste and microplastics in coastal ecosystems. Galgani's citations peaked and became prominent as foundational work in monitoring sea plastic debris in most of the world; Honorato-Zimmer and Chen highlighted emerging relevant sectors, namely citizen science and an approach toward regional pollution assessments.

Based on the temporal variation, a second pattern of research focus is symbolized with diversification, as microplastics and policy solutions emerged as research subjects in addition to methods, as seen in Barcelo and Damar These citation metrics contrast with Guo and Honorato-Zimmer's lower outputs, highlighting these are low-reach but high-impact studies. Overall, Fig. 5.9 showcases a field motivated by seasoned leaders and up-and-coming scholars, with collaboration and interdisciplinary pathways (e.g., connecting tourism with ecology) emerging as cornerstones to tackle the global challenges posed by plastic pollution.

Table 5.3 provides insights into the citation dynamics of these foundational studies on plastic pollution with global or local reach. Napper *et al.* (2020), which has 406 global citations and a low LC/GC ratio (0.99%), features research that has a global impact, which we suspect is a consequence of its focus on microplastics in extreme environments, such as in Mount Everest, which has relevance in multiple national contexts. In contrast, Jang *et al.* (2014) accordingly have a high local relevance (LC/GC ratio: 3.96%), as the research localizes its findings on marine debris effects on the tourism economy of Geoje Island in South Korea. Similarly, although old, Santos *et al.* (2005) still present a middle LC/GC ratio (2.11%), which indicates the continuing local importance of these studies, allowing us to understand the beach litter generation when correlated to socioeconomic aspects in Brazil.

Willis *et al.* (2018) and Retama *et al.* (2016) have dissimilar trends: Willis *et al.*'s policy analysis on marine plastics receives greater average normalized global citations (1.87) than Retama *et al.* moderates local (4) with international focus (129 citations) Honorato-Zimmer *et al.* (2019) features a very high LC/GC ratio (7.41%), accompanied by increased normalized local citations (4), underlining the importance of their citizen science methodologies in the geographical context of Chile and Germany.

Recent work, such as Gosavi and Phuge (2023), shows emerging local traction (LC/GC ratio: 9.09%) but with limited worldwide citations (11) given their novelty, and Becherucci and Seco Pon (2014) whose niche influence is little felt locally despite some interest globally. The data highlights a dual trend: studies oriented toward global impacts (e.g., microplastics in pristine ecosystems) dominate citation metrics, while region-specific work (e.g., pollution in coastal tourism centroids) catalyzes localized action in academia and policymaking. Normalized citations show that newer studies (e.g., Gosavi and Phuge, 2023) are gaining local relevance at an increasing rate, indicating an increased focus on solutions specific to that context. A table showing the interplay between global and local research in tackling plastic pollution.

Figure 5.10 shows the keyword co-occurrence in the plastic pollution research based on the bibliometric network using VOSviewer in the keywords space based on the explosion of resources used as a technical concept associated with microplastics, marine pollution and waste management due to credible journal selection. The faces of the brilliant bunches are the thematic regions as per the settlement of research work.

The green cluster generally relates to pollution in the marine environment, waste disposal, and environmental pollution. Phrases such as 'marine pollution' and 'waste management', 'beach' and 'seashore' indicate a focus on the environmental effects of plastic debris on coastal and marine environments. The blue cluster encompasses the human aspect of plastic pollution, with keywords such as humans, environmental protection, and ecosystem, suggesting research addressing the socioeconomic and ecological interactions with pollution. This cluster also includes neural pathways related to climate change, biodiversity, and population

Table 5.3. Citation dynamics of plastic contamination foundational studies. Author's own.

Document	DOI	Year	Local citations	Global citations	LC/GC Ratio (%)	Normalized local citations	Normalized global citations
Napper et al., 2020, ONE EARTH	10.1016/j.oneear.2020.10.020	2020	4	406	0.99	13	1.01
Laglbauer et al., 2014, MAR POLLUT BULL	10.1016/j.marpolbul.2014.09.036	2014	4	373	1.07	0.92	0.99
Jang et al., 2014, MAR POLLUT BULL	10.1016/j.marpolbul.2014.02.021	2014	8	202	3.96	1.85	4.1
Santos et al., 2005, OCEAN COAST MANAGE	10.1016/j.ocecoaman.2005.08.006	2005	4	190	2.11	1	0.87
Willis et al., 2018, MAR POLICY	10.1016/j.marpol.2017.11.037	2018	1	140	0.71	2	1.87
Retama et al., 2016, MAR POLLUT BULL	10.1016/j.marpolbul.2016.08.053	2016	4	129	3.1	2.4	1
Honorato-Zimmer et al., 2019, MAR POLLUT BULL	10.1016/j.marpolbul.2018.11.048	2019	4	54	7.41	4	1.11
Becherucci and Seco Pon, 2014, WASTE MANAGE	10.1016/j.wasman.2014.02.020	2014	1	23	4.35	0.23	0.55
Gosavi and Phuge, 2023, ENVIRON SCI POLLUT RES	10.1007/s11356-023-27074-2	2023	1	11	9.09	14	0.12

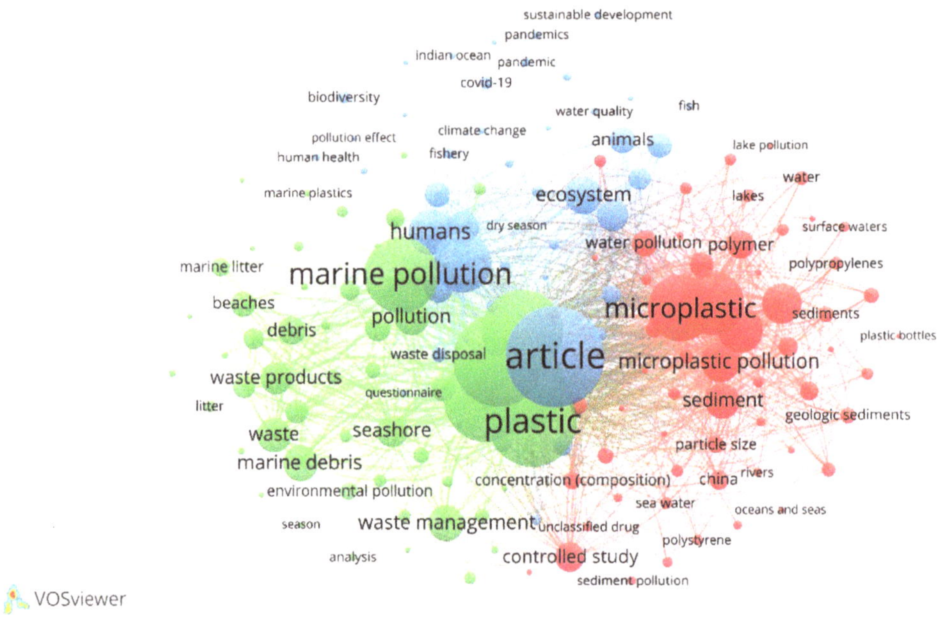

Fig. 5.10. Keyword co-occurrence. Author's own.

dynamics, which are important drivers of environmental protection efforts.

However, the red cluster is highly relevant to microplastic, water, and sediment pollution. The predominant keywords such as 'microplastic', 'water pollutant', 'polymer', and 'sediment' suggest the research is concentrated on its chemical and physical properties, distribution, and effects on aquatic receivers. Therefore, these tightly coupled keywords indicate much research exists on microplastic pathways and environmental persistence.

Figure 5.10 reveals the interconnectivity of the research themes in plastic pollution studies and reflects the multidisciplinary context of this field. The emergence of keywords highlights major research domains, including environmental implications, socioeconomic aspects, and pollution reduction strategies. The overall state of research on plastic pollution presented in this bibliometric analysis highlights the structure this research is currently taking and illustrates where further investigations may be warranted to fill gaps in knowledge.

This keyword cloud visualization provides a view of the key themes and most frequent occurring terms in this research domain of environmental monitoring and pollution (Fig. 5.11). The most salient term, 'environmental monitoring' suggests a prevalent concern and drive to monitor and understand environmental changes, especially regarding pollution. Other relevant terms are 'plastic waste', 'marine pollution', 'microplastic', and 'ecosystem', revealing a growing recognition of plastic pollution in marine and terrestrial systems.

Keywords like 'tourism', 'waste management', 'water pollutants', and 'human activities' indicate anthropogenic causes of environmental degradation. Its word cloud also shows geographical and sector-specific mentions like 'China', 'India', and 'Mediterranean Sea', representing region-centric research on pollution.

This word cloud resonates visually with the close relationship between the category of environmental issues, in particular highlighting plastic pollution, the health of the ecosystem and sustainability challenges related to industrial, agricultural, energy, etc. It is an informative article and reminds us that environmental risks, waste management, and what we can do to protect natural ecosystems are rapidly becoming the stuff of academic and policy debate.

Fig. 5.11. Word cloud. Author's own.

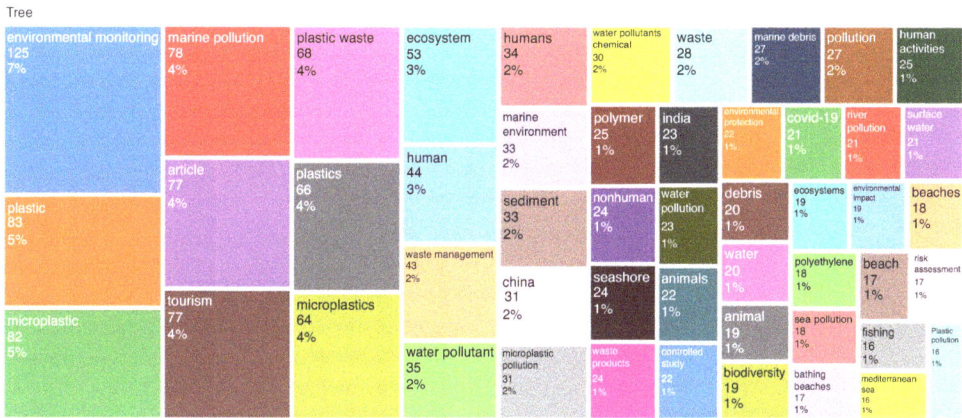

Fig. 5.12. Tree map. Author's own.

This tree map visualization (Fig. 5.12) illustrates the relative importance of the top terms relating to environmental monitoring, pollution, and sustainability within the literature. In particular, 'environmental monitoring' (125, 7%) is the most pronounced keyword, emphasizing the importance of monitoring environmental change and pollution levels. Noteworthy terms and their respective frequencies include 'plastic' (83, 5%), 'microplastic' (82, 5%), and 'marine pollution' (78, 4%), implying that global research works on plastic pollution (e.g., microplastic) and marine ecosystems are elaborately focused on.

Similarly, 'waste management' (43, 2%) and 'water pollutant' (35, 2%) underscore concerns of waste disposal and contamination, while 'ecosystem' (53, 3%) and 'biodiversity' (19, 1%) suggest attention to the maintenance of ecological equilibrium. 'Tourism' (77, 4%) and 'beaches' (18, 1%) hint at the connection between human activities and the degradation of the environment, especially by the water.

Other emerging topics like 'covid-19' (21, 1%) and 'human activities' (25, 1%) indicate recent hot topics on various developments in environmental studies that mainly focus on the influence of the pandemic on the pattern of pollution. In conclusion, the visualization highlights the multifaceted nature of plastic pollution and its far-reaching impact on human health, ecosystems, and the environment at

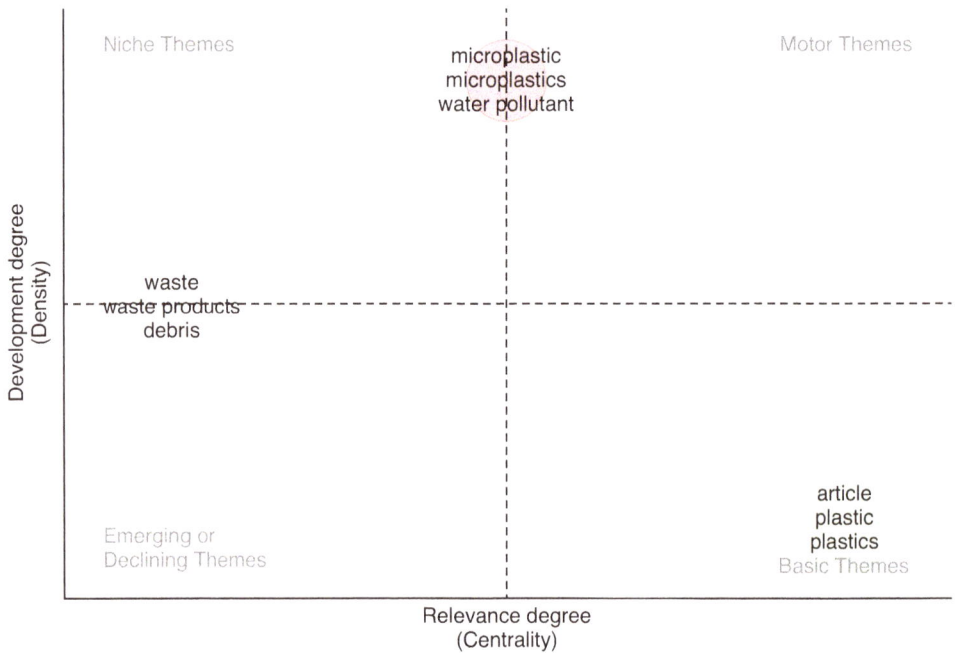

Fig. 5.13. Thematic analysis. Author's own.

large, emphasizing the need for continued research efforts in this field.

Figure 5.13 shows a thematic analysis of the literature about plastic pollution according to the categories for each theme as a function of its development (maturity) and relevance (centrality). The quadrants are niche themes, motor themes, emerging/declining themes, and basic themes. Motor themes are mature and key to the research domain. Microplastics and microplastics represent important key themes in this quadrant of the bibliographic coupling network, illustrating its relevance to current studies. They are critical for grappling with the ubiquity of microscopic plastic debris on ecosystems and human health. 'Water pollutant' is also visible in this quadrant, signifying the need to address water to plastic waste pollution. The niche themes are more fleshed out and less exploratory. Themes such as 'waste', 'waste products' and 'debris' come under this head.

While important, these areas may not be cited as often or melded into larger discussions of research as motor themes or critical themes. Essential and central but less-developed basic themes These topics are the basis of the research

area and can be further developed. The certain quadrants in this diagram do not include the actual subjects. Yet, reasonable topics would usually be basic ideas that everybody may be aware of but have not occurred. Emerging or declining themes are newer and less central. They are new areas of interest or seeking relevance. The image indicates that some trends associated with plastic pollution may be coming into play as new research priorities, and others might be waning. As such, this thematic analysis presents an organized overview of the field of exploration for research that highlights areas that are already well established, areas that are emerging and areas of research that may need more focus in the future. This framework may help set a direction for future research efforts to identify what is currently lacking in the plastic pollution research field.

The integration of this information in the form of world map visualization can indicate any geographical distribution or collaboration networks in environmental research data, where different shades of blue indicate where most studies are done (Fig. 5.14). The darker the color the more research contribution, and

Country Collaboration Map

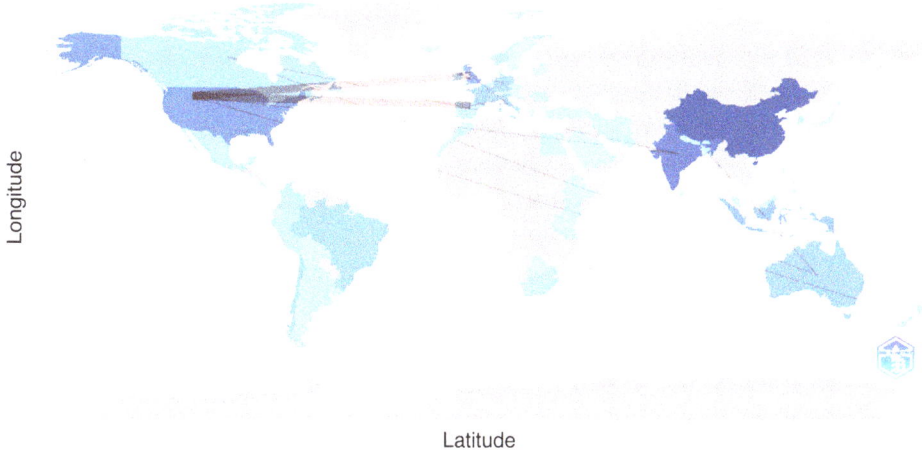

Fig. 5.14. World map. Author's own.

the lighter the color the less contribution. The set of distinguished environmental researchers were also presented, coming primarily from the USA, China, and India are the regions that stand out most in quantity with respect to the topics of study, suggesting that these regions continue to play a prominent role in environmental research focused on pollution, sustainability, and climate change. The North Atlantic Oscillation (NAO) offers a strong research engagement of Canada, Australia, and many European nations. Research collaborations also appear on the map, as connecting lines that show strong international partnerships, especially for North America, Europe, and Asia. These connections highlight the international aspect of environmental research, with nations collaborating on topics such as plastic pollution, climate change, and development of ecosystems.

The high emissions from Asia, particularly China and India, indicate an increasing interest in environmental issues in the fast-industrializing world. Similarly, the US and Transatlantic countries have been part of this international effort led by the work they already do in environmental research and policy. Australia, with its prominent role in the network, points to its contributions, probably around marine and climate, given the prominence of work in those areas. This visualization highlights the scale of global

collaboration needed to address these environmental challenges and the impact of regional research nodes with the United States, China, India, and Europe as regional leaders driving the sustainability and conservation discourse.

Figure 5.15 shows the trend topics in environmental research from 2014 to 2024, revealing the progression of key topics through the years. The size of each bubble corresponds to term frequency, indicating how prominent the respective topics are in research papers.

The analysis showed that early research trends (2014–2018) were characterized as 'environmental pollution', 'comparative study', and 'waste management', which created the basis of subsequent studies. A few years later, in 2018, even more specialized topics started to arise: 'marine pollution', 'plastic' and 'environmental monitoring', reflecting concerns about contamination of oceanic and atmospheric environments. The terms relevant to microplastics, such as microplastics, microplastic pollution and plastic, show excellent growth from 2020 to 2024. Regional studies' recommendations for pollution and environmental sustainability are indicated in the frequent mention of geographical locations such as 'China', 'Chile', and 'Argentina'. The emergence of new keywords such as 'tourism', 'beaches', 'recreation', and 'bathing beaches' suggests that

Trend Topics

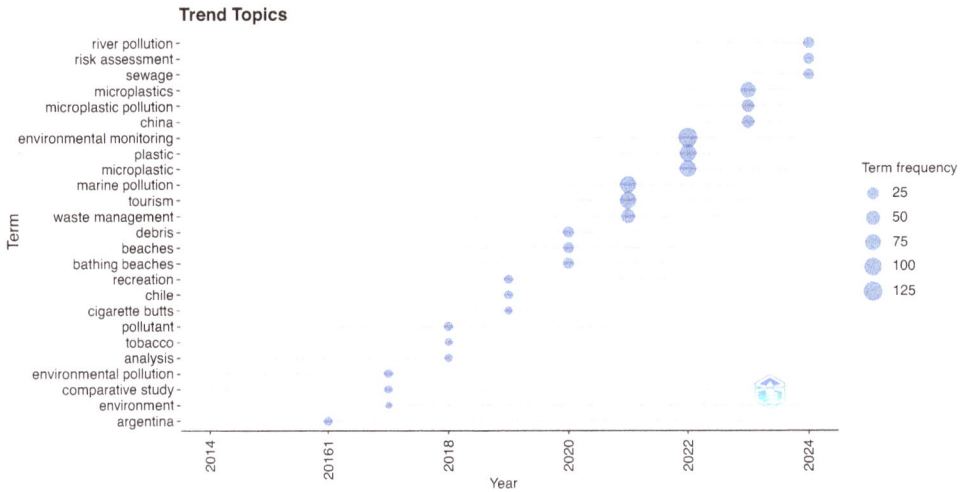

Fig. 5.15. Trend topics. Author's own.

we are capturing this economic–environmental intersection, especially in the context of coastal areas. The latest trend (2022–2024) focuses on risk assessment, cigarette butts as pollutants, and when it comes to specific water pollution issues, 'river pollution' and 'sewage' are the main keywords. This reflects a paradigm shift in which an increasing range of pollution mitigation approaches are being implemented, along with assessing the environmental impact of human actions across the landscape. Figure 5.15 shows the gradual refinement of environmental analysis; we go from general topics of pollution to very particular issues related to waste management and microplastic contamination.

5.5 Findings

The bibliometric analysis investigates existing literature on plastic pollution in the context of tourism. As depicted in Fig. 5.1, there has been a significant increase in publications after 2014 associated with a growing awareness of the impact of tourism on plastic waste and the ecological effects thereof. Since 2016, Fig. 5.2 shows peaks of citations per year corresponding to (i) global environmental campaigns around 2016 and (ii) a resurgence in publications post-2018, when a sense of urgency around addressing the impacts of microplastics rose, as did a

wider recognition of a hitherto bungled policy response to those materials. The compounding effects of keyword analysis underlined the prevalence of eco-threat emphasis on 'microplastic' and 'plastic' while incorporating words of value, such as 'marine pollution', and linking terms, such as 'tourism' suggesting sectoral issues. The limited reference to 'environmental monitoring' points toward a need for more robust methodological arsenals to document pollution.

Internationally, China contributes the highest number of research papers on plastic pollution, followed by the USA and India, each representing developed and developing nations, thus showcasing the collective concern on a global scale. Such examples are evident in regional studies. For instance, Jang *et al.* (2014) study on South Korea's Geoje Island shows localized impacts, whereas foundational studies such as Napper *et al.* (2020) are on Mount Everest, and Forster *et al.* (2023) on global significance. *Marine Pollution Bulletin* and *Science of the Total Environment* are identified as core journals by Bradford's Law, as the journals strongly influence the spread of high-impact research. A plot of citation dynamics indicates two areas of interest – papers with any global application (e.g., microplastics in extreme environments) and those with regional relevance (e.g., citizen science in Chile).

Thematic clusters include motor themes (e.g., impacts of microplastics, water pollution), niche themes (e.g., waste management), and emerging trends (e.g., COVID-19 PPE waste), reflecting a multidisciplinary field. Geographic mapping using longitude/latitude data enables pollution hot-spotting (e.g., coastal tourism centers) and trend analysis from 2014 to 2024 demonstrates a transition from broad pollution subjects to targeted foci such as microplastic risk assessment and regional waste management strategies. Collaborative authorship patterns (6.04 coauthors per document) and pan-global partnerships (32.46%) are a hallmark of foreign direct investment, a stark contrast to the national and federal scale of the Canada–US relationship, where the impacts of transboundary environmental issues are felt, and the complexity of dealing with them is apparent.

Balancing solid science and policies in practice, particularly circular economy models, alternatives and degradable substitutes and tourism practices in line with the UN Global Sustainable Development Goals (SDGs). Future studies need to reach underrepresented biomes (e.g., inland freshwater systems) and enhance monitoring frameworks to reduce the growing threat of plastic pollution to ecosystems and human well-being.

5.6 Implications of the Study

5.6.1 Theoretical implications

Bibliometric analysis provides important theoretical contributions by mapping plastic pollution research evolution in the tourism field, indicating a diagonal shift from macro to micro scale studies. This highlights the importance of developing interdisciplinary frameworks that better address the multifaceted relationship between tourism and plastic waste, considering ecological, socioeconomic, and policy aspects. The overall frequency and use of terms like 'microplastic' and 'marine pollution' in recent studies suggests that comprehensive knowledge of cumulative and long-term ecological impacts remains largely unknown, particularly in less examined settings like inland freshwater systems and

mountainous areas. Moreover, trends are observed on both global SC/GC ratios (e.g., Napper et al., 2020) and localized LC/GC ratios (e.g., Jang et al., 2014), which also bring a perspective for developing context-sensitive theories that combine universal environmental principles with local sociocultural dynamics. In the future, this could become an important focus of research as plastic categories with predictive models have been created to estimate cumulative plastic leakage, and the methodology for real-time pollution monitoring needs further refinement. This will also provide theoretical rigor to sustainability science.

5.6.2 Managerial implications

Policymakers and industry stakeholders are illustrated by the results of actionable strategies in addressing tourism's plastic footprint. Research shows that coastal and marine pollution prevails; thus, targeted waste management policies are needed: a ban on single-use plastics in tourist centers, for instance; however, give incentives for circular economy solutions (e.g., reusable garbage). Regional studies' relatively high LC/GC ratios (e.g., Honorato-Zimmer et al., 2019) argue for decentralized governance where the local community shapes citizen science activities or rules and regulations. Develop tourism operators to take up such certification schemes that promote plastic-free practices at tourist spots; spatial data (longitude/latitude mapping) can help governments identify high-impact zones for cleanup and infrastructure upgrades. Academia, industry, and NGOs must work closely together, as reflected by the 32.46% international coauthorship scaled up to share best practices and push the technologies necessary to help world recovery. This sustainable tourism development policy should be integrated into national sustainable development actions for responsible consumption (SDG 12) and below-water ecosystem conservation (SDG 14) as guided by research to achieve a sustainable coexistence of tourism development and environmental sustainability.

5.7 Future Research Directions

5.7.1 Emerging topics in plastic pollution and tourism

Newer studies have begun to showcase different perspectives on the relationship between plastic pollution and tourism, describing how microplastics damage marine environments and could damage tourists' experience. People are critical in driving consumption patterns and must implement sustainable habits to reduce plastic waste. The implications of such emerging technologies, capable of delivering massive cuts in plastic waste, are important for sustainable tourism policies. A lack of community awareness of plastic pollution because of the prevalence of plastic-themed sustainability marketing of tourism packages, especially with the sustainability 'trend', has also been suggested as a common concern. Case studies from known tourist spots that have successfully put targeted and tailored interventions into place establish a direct connection between decreased plastic pollution and increased tourist satisfaction. One of the areas that can contribute to the future of tourism-related products is the study of biodegradable materials.

5.7.2 Interdisciplinary approaches to addressing plastic pollution

Not only does this multidisciplinary approach, which weaves together environmental science, tourism studies and sociology, yield more holistic solutions, it enables academia, industry, and policymakers to partner. Technological innovations have been central in how plastic waste can be tracked and reduced, and local communities have been necessary for conservation efforts and educational outreach. These multidisciplinary projects have a proven pedigree in successful case studies from tourism hotspots. There is a useful lash here

on changing tourist behavior, but it is also important to have strong policy frameworks to encourage sustainable behavioral practices in the tourism sector.

5.7.3 The role of technology in mitigating plastic pollution

Technology disposal is one of the ways to combat the plastic pollution crisis. Biodegradable materials and single-use plastic substitutes used by stakeholders can reduce the environmental footprint significantly. Intelligent waste management systems optimize recycling, and mobile apps encourage visitors to engage in sustainable behavior. Leading the way in this regard are tech companies that help monitor and report levels of plastic pollution in those areas and work with those involved in the tourism experience to create sustainable solutions. Furthermore, artificial intelligence also saves space and reduces plastic waste – hence, educating tourists on plastic pollution should be enhanced through digital platforms.

5.7.4 Recommendations for future studies

This chapter calls for future research on the more detailed interaction between plastic pollution and local economies of tourist destinations because it is important to comprehend that interaction to have sustainable development. Also, a comprehensive examination of the effect existing policies and regulations in tourism have on mitigating plastic waste is necessary. Long-term convergence research into changing plastic pollution in popular local tourism locations over defined periods can provide key insight. Further investigation of the effects of information and awareness campaigns on tourist behaviors and examining evidence of effective plastic reduction initiatives would also benefit the debate. By studying the viability of replacements for plastic products, the working relationship between tourism companies and local populations and monitoring plastic waste through tech, we can continue to develop a greater understanding of this urgently important topic. Finally, investigating tourists' perceptions and attitudes regarding plastic pollution and sustainability can help shape more impactful change initiatives.

5.8 Conclusion

This bibliometric analysis illustrated the vital role of plastic pollution research in tourism. These findings prompt the growing interest in the field of research on sustainable development and tourism management and highlight the need for the tourism sector to respond to the increasing volume of plastic waste. Keywords used for the review process included pellets, environmental degradation, economic burden, social issues, and tourism impacts. It underscores stakeholders' need to adopt sustainable practices that reduce plastic pollution, conserve natural resources and enhance the tourist experience. Furthermore, our bibliometric analysis utilized a methodology that demonstrates a solid frame for the research navigation trails of existing work regarding, and gaps within, plastic pollution in tourism, providing the cornerstone for future studies that can reveal the complexities of plastic pollution in tourism even more. The commentary on the findings illustrates the promise of interdisciplinary approaches combining predictive information from environmental science, economics, and issues within the tourism field to create something new. Future efforts should focus on policies and practices that will have the greatest impact in addressing these plastics. This is a win–win situation where we not only save our ecosystem but also secure the long-term viability of responsible tourism, and we work toward a better economy and a healthy earth for generations to come.

References

Aragaw, T.A., De-la-Torre, G.E. and Teshager, A.A. (2022) Personal protective equipment (PPE) pollution driven by the COVID-19 pandemic along the shoreline of Lake Tana, Bahir Dar, Ethiopia. *Science of the Total Environment* 820, 153261. DOI: 10.1016/j.scitotenv.2022.153261.

Becherucci, M.E. and Seco Pon, J.P. (2014) What is left behind when the lights go off? Comparing the abundance and composition of litter in urban areas with different intensity of nightlife use in Mar del Plata, Argentina. *Waste Management* 34(8), 1351–1355.

Carlsson, P., Singdahl-Larsen, C. and Lusher, A.L. (2021) Understanding the occurrence and fate of microplastics in coastal arctic ecosystems: The case of surface waters, sediments and walrus (*Odobenus rosmarus*). *Science of the Total Environment* 795, 148308. DOI: 10.1016/j.scitotenv.2021.148308.

Dube, K. and Nhamo, G. (2021) Sustainable development goals localisation in the tourism sector: Lessons from Grootbos Private Nature Reserve, South Africa. *GeoJournal* 86(6), 2831–2846. DOI: 10.1007/s10708-020-10182-8.

Forster, P.M., Smith, C.J., Walsh, T., Lamb, W.F., Lamboll, R. *et al.* (2023) Indicators of global climate change 2022: Annual update of large-scale indicators of the state of the climate system and human influence. *Earth System Science Data* 15(6), 2295–2327.

Garcés-Ordóñez, O., Espinosa Díaz, L.F., Pereira Cardoso, R. and Costa Muniz, M. (2020) The impact of tourism on marine litter pollution on Santa Marta beaches, Colombian Caribbean. *Marine Pollution Bulletin* 158, 111558. DOI: 10.1016/j.marpolbul.2020.111558.

Gosavi, S.M. and Phuge, S.K. (2023) First report on microplastics contamination in a meteorite impact crater lake from India. *Environmental Science and Pollution Research* 30(23), 64755–64770.

Gregory, M.R. (1999) Plastics and South Pacific island shores: Environmental implications. *Ocean and Coastal Management* 42(6–7), 603–615. DOI: 10.1016/S0964-5691(99)00036-8.

Honorato-Zimmer, D., Kruse, K., Knickmeier, K., Weinmann, A., Hinojosa, I.A. *et al.* (2019) Interhemispherical shoreline surveys of anthropogenic marine debris – a binational citizen science project with schoolchildren. *Marine Pollution Bulletin* 138, 464–468. DOI: 10.1016/j.marpolbul.2018.11.048.

Jang, Y.C., Hong, S., Lee, J., Lee, M.J. and Shim, W.J. (2014) Estimation of lost tourism revenue in Geoje Island from the 2011 marine debris pollution event in South Korea. *Marine Pollution Bulletin* 85(1), 303–306. DOI: 10.1016/j.marpolbul.2014.02.021.

Krishnakumar, S., Anbalagan, S., Kasilingam, K., Smrithi, P., Anbazhagi, S. *et al.* (2020) Assessment of plastic debris in remote islands of the Andaman and Nicobar Archipelago, India. *Marine Pollution Bulletin* 151, 110841. DOI: 10.1016/j.marpolbul.2019.110841.

Laglbauer, B.J.L., Franco-Santos, R.M., Andreu-Cazenave, M., Brunelli, L., Papadatou, M. *et al.* (2014) Macrodebris and microplastics from beaches in Slovenia. *Marine Pollution Bulletin* 89(1–2), 356–366. DOI: 10.1016/j.marpolbul.2014.09.036.

Laju, R.L., Jayanthi, M., Jeyasanta, K.I., Patterson, J., Asir, N.G.G. *et al.* (2022) Spatial and vertical distribution of microplastics and their ecological risk in an Indian freshwater lake ecosystem. *Science of the Total Environment* 829, 153337. DOI: 10.1016/j.scitotenv.2022.153337.

Lim, Y.C., Chen, C.-W., Cheng, Y.-R., Chen, C.-F. and Dong, C.-D. (2022) Impacts of microplastics on scleractinian corals nearshore Liuqiu Island Southwestern Taiwan. *Environmental Pollution* 307, 119371. DOI: 10.1016/j.envpol.2022.119371.

MacLeod, M., Arp, H.P.H., Tekman, M.B. and Jahnke, A. (2021) The global threat from plastic pollution. *Science* 373(6550), 61–65. DOI: 10.1126/SCIENCE.ABG5433.

Mofijur, M., Ahmed, S.F., Rahman, S.M.A., Siddiki, S.Y.A., Islam, A.B.M.S. *et al.* (2021) Source, distribution and emerging threat of micro- and nanoplastics to marine organism and human health: Socio-economic impact and management strategies. *Environmental Research* 199, 110857. DOI: 10.1016/j.envres.2021.110857.

Nachite, D., Maziane, F., Anfuso, G. and Williams, A.T. (2019) Spatial and temporal variations of litter at the Mediterranean beaches of Morocco mainly due to beach users. *Ocean and Coastal Management* 181, 104846. DOI: 10.1016/j.ocecoaman.2019.104846.

Napper, I.E., Davies, B.F.R., Clifford, H., Elvin, S., Koldewey, H.J. *et al.* (2020) Reaching new heights in plastic pollution – preliminary findings of microplastics on Mount Everest. *One Earth* 3(5), 621–630. DOI: 10.1016/j.oneear.2020.10.020.

Ormaza-González, F.I., Castro-Rodas, D. and Statham, P.J. (2021) COVID-19 impacts on beaches and coastal water pollution at selected sites in Ecuador, and management proposals post-pandemic. *Frontiers in Marine Science* 8, 669374. DOI: 10.3389/fmars.2021.669374.

Patterson Edward, J.K., Jayanthi, M., Malleshappa, H., Immaculate Jeyasanta, K., Laju, R.L. *et al.* (2021) COVID-19 lockdown improved the health of coastal environment and enhanced the population of reef-fish. *Marine Pollution Bulletin* 171, 112124. DOI: 10.1016/j.marpolbul.2021.112124.

Qiang, M., Shen, M. and Xie, H. (2020) Loss of tourism revenue induced by coastal environmental pollution: A length-of-stay perspective. *Journal of Sustainable Tourism* 28(7), 1025–1043. DOI: 10.1080/09669582.2019.1684931.

Quayle, D.V. (1992) Plastics in the marine environment: Problems and solutions. *Chemistry and Ecology* 6, 69–78. DOI: 10.1080/02757549208035263.

Reed, S., Clark, M., Thompson, R. and Hughes, K.A. (2018) Microplastics in marine sediments near Rothera Research Station, Antarctica. *Marine Pollution Bulletin* 133, 460–463. DOI: 10.1016/j.marpolbul.2018.05.068.

Retama, I., Jonathan, M.P., Shruti, V.C., Velumani, S., Sarkar, S.K. *et al.* (2016) Microplastics in tourist beaches of Huatulco Bay, Pacific Coast of Southern Mexico. *Marine Pollution Bulletin* 113(1–2), 530–535. DOI: 10.1016/j.marpolbul.2016.08.053.

Santos, I.R., Friedrich, A.C., Wallner-Kersanach, M. and Fillmann, G. (2005) Influence of socio-economic characteristics of beach users on litter generation. *Ocean and Coastal Management* 48(9–10), 742–752. DOI: 10.1016/j.ocecoaman.2005.08.006.

Semernya, L., Ramola, A., Alfthan, B. and Giacovelli, C. (2017) Waste management outlook for mountain regions: Sources and solutions. *Waste Management and Research* 35(9), 935–943. DOI: 10.1177/0734242X17709910.

Shu, X., Xu, L., Yang, M., Qin, Z., Zhang, Q. *et al.* (2023) Spatial distribution characteristics and migration of microplastics in surface water, groundwater and sediment in Karst areas: The case of Yulong River in Guilin, Southwest China. *Science of the Total Environment* 876, 161578. DOI: 10.1016/j.scitotenv.2023.161578.

Suteja, Y., Atmadipoera, A.S., Riani, E., Nurjaya, I.W., Nugroho, D. *et al.* (2021) Stranded marine debris on the touristic beaches in the south of Bali Island, Indonesia: The spatiotemporal abundance and characteristic. *Marine Pollution Bulletin* 173(Pt A), 113026. DOI: 10.1016/j.marpolbul.2021.113026.

Wabnitz, C.C.C. and Nichols, W.J. (2010) Editorial. Plastic pollution: An ocean emergency. *Marine Turtle Newsletter* 129, 1–4.

Watt, E., Picard, M., Maldonado, B., Abdelwahab, M.A., Mielewski, D.F. *et al.* (2021) Ocean plastics: Environmental implications and potential routes for mitigation – a perspective. *RSC Advances* 11(35), 21447–21462. DOI: 10.1039/D1RA00353D.

Willis, K., Maureaud, C., Wilcox, C. and Hardesty, B.D. (2018) How successful are waste abatement campaigns and government policies at reducing plastic waste into the marine environment. *Marine Policy* 87, 94–101. DOI: 10.1016/j.marpol.2017.11.037.

6 Managing the Multitude: A Study of Overcrowding and Sustainable Practices at the Char Dham Pilgrimage Sites

Om Sharma* and Alka Maheshwari
Amity University, India

Abstract

Char Dham in Uttarakhand is a significant pilgrimage site attracting millions of devotees each year, deeply rooted in spiritual and cultural traditions. However, the growing influx of visitors has led to overcrowding, environmental degradation, and pressure on local infrastructure. These challenges have strained natural resources, disrupted the sanctity of the pilgrimage, and impacted local communities. Although sustainable tourism measures have been introduced, enforcement gaps and inadequate planning have hindered their effectiveness. This chapter examines the consequences of mass tourism on Char Dham and explores solutions such as visitor management, infrastructural development, environmental conservation, and community-led initiatives. By adopting these strategies, the pilgrimage can remain a sacred and fulfilling experience while preserving the region's fragile ecosystem for future generations.

6.1 Introduction

For centuries, the Char Dham Yatra has been one of Hinduism's most revered pilgrimages, drawing millions of devotees to the sacred shrines of Yamunotri, Gangotri, Kedarnath, and Badrinath in Uttarakhand (Singh and Rana, 2023). Traditionally, this journey was a test of faith and endurance, demanding deep spiritual reflection. However, modernization has transformed the pilgrimage, making it more accessible through improved roads, helicopter services, and digital promotions (Mohanty and Mishra, 2021; Chandan *et al.*, 2023). What was once an arduous, introspective experience has become, for many, a rushed trip through crowds, logistical challenges, and increasing commercialization (Talukder *et al.*, 2024).

The rapid surge in visitors has placed immense pressure on both the pilgrimage and the fragile Himalayan ecosystem. Long queues, overcrowding, and noise disrupt moments of spiritual solitude, while unchecked tourism contributes to pollution, deforestation, and the deterioration of vital pilgrimage routes (Kowalik *et al.*, 2022; Khalilov, 2023; Shashwat *et al.*, 2024). Local infrastructure struggles to support the influx, leaving pilgrims – especially the elderly – facing issues such as inadequate sanitation and limited medical aid (Talukder *et al.*, 2024).

This study examines the impact of overcrowding on the Char Dham experience, environment, and local communities. By analyzing sustainable models from other religious sites and applying pilgrimage management theories,

*Corresponding author: om.sha.24071@gmail.com

© CAB International 2025. *Municipal Waste Management: Policies and Strategies*
(eds S. Kumar *et al.*)
DOI: 10.1079/9781836990666.0006

it explores ways to preserve the sanctity of Char Dham while ensuring its accessibility (Khalilov, 2023; Yadav, 2024). A pilgrimage should be a journey of faith, not frustration. With thoughtful policies, technology, and community-driven initiatives, it is possible to harmonize devotion with sustainability, safeguarding this sacred tradition for generations to come (Saeedi, 2022).

6.2 Objectives

This research aims to explore the complex challenges posed by overcrowding at the Char Dham pilgrimage sites while identifying sustainable solutions that enhance the overall pilgrimage experience without compromising the ecological and cultural integrity of these revered destinations. By adopting a structured approach, the study seeks to address three key objectives, each of which examines a critical aspect of pilgrimage tourism and sustainability.

1. To analyze pilgrim behavior patterns, preferences, and the impact of overcrowding on pilgrimage satisfaction based on collected data.
2. To compare the effectiveness of existing sustainable practices at the Char Dham sites with those implemented at similar pilgrimage destinations globally.
3. To develop actionable recommendations for managing overcrowding and promoting sustainable tourism at the Char Dham sites based on the analyzed data.

6.3 Literature Review

Pilgrimage tourism has long been studied in relation to sustainability, religious significance, and socioeconomic impact. Scholars highlight that sacred sites worldwide face common challenges – overcrowding, commercialization, and environmental degradation (Yadav, 2024). To address these issues, sustainable tourism models, such as the Global Sustainable Tourism Council framework, emphasize balancing environmental protection, visitor management, and

infrastructure planning to preserve the sanctity of pilgrimage sites (Khalilov, 2023).

Overcrowding remains a pressing issue at major Indian pilgrimage destinations like Varanasi, Mathura, Tirupati, Puri, and Haridwar, where unchecked footfall has led to resource depletion, excessive waste, and tensions between locals and visitors (Sharma, 2020; Mohanty and Mishra, 2021). The strain on these sacred places highlights the urgent need for better visitor management policies that protect their cultural and spiritual integrity. Similarly, international religious sites like Mecca, the Vatican, and the Camino de Santiago have adopted structured pilgrimage management strategies, such as regulated entry systems and sustainable tourism frameworks, to mitigate the adverse effects of mass tourism.

The environmental consequences of religious tourism are especially severe in ecologically fragile regions like the Himalayas. The expansion of roads, hotels, and parking facilities to accommodate rising pilgrim numbers has led to deforestation, habitat destruction, and an increased risk of landslides (Kowalik et al., 2022; Shashwat et al., 2024). While improved access enhances the pilgrimage experience, it raises concerns about long-term ecological damage and the sustainability of traditional routes.

Beyond environmental challenges, the economic and cultural dimensions of pilgrimage tourism require thoughtful management. While pilgrimage-driven tourism generates income for local businesses, it also risks reducing sacred rituals to commercial transactions, eroding the authenticity of religious experiences (Singh and Gowreesunkar, 2019). Scholars suggest that community involvement in conservation and tourism planning can help protect cultural heritage while ensuring that economic benefits reach local residents (Saeedi, 2022).

Technology has emerged as a potential solution for managing large crowds and improving pilgrimage experiences. Innovations like e-ticketing, real-time crowd monitoring, and eco-friendly transport have been introduced in major religious destinations (Arora et al., 2023). However, implementing such solutions in remote Himalayan areas like Char Dham remains challenging due to logistical constraints and limited digital adoption. The effectiveness of smart tourism depends not only on infrastructure but also on collaboration between stakeholders willing to integrate modern solutions into traditional pilgrimage management.

This review underscores the need for a structured, sustainable approach to managing pilgrimage tourism. By adopting recognized sustainability frameworks, engaging local communities, and leveraging technology-driven solutions, religious tourism can be regulated to ensure that sacred sites retain their spiritual essence while contributing to regional development. With responsible policies and environmental stewardship, pilgrimage destinations can continue to inspire faith without being overwhelmed by the pressures of mass tourism.

6.4   Research Methodology

This study employs a mixed methods approach, combining both quantitative and qualitative research to evaluate the impact of overcrowding at Char Dham. Structured surveys conducted with pilgrims offer firsthand insights into their experiences, while secondary data from academic studies, government reports, and global pilgrimage case studies provide a broader contextual understanding. By analyzing visitor experiences, infrastructure challenges, environmental concerns, and community impacts, the study highlights key issues and explores practical, sustainable solutions. This comprehensive methodology ensures that the proposed recommendations not only enhance management strategies but also preserve the religious and cultural significance of Char Dham.

6.5   Research Design

The research design is structured to examine three fundamental aspects of pilgrimage sustainability at Char Dham:

1. *Pilgrim experience and satisfaction*: analyzing how overcrowding affects the overall spiritual journey, including access to temple rituals, waiting times, and visitor engagement in religious activities.
2. *Environmental and infrastructural challenges*: assessing the strain on local ecosystems,

waste accumulation, the availability of essential public amenities such as sanitation and drinking water, and the resilience of transport and accommodation infrastructure.
3. *Sustainable tourism strategies*: evaluating best practices in pilgrimage site management from international case studies and identifying potential strategies for implementing visitor control, ecological conservation, and transport management in the Char Dham circuit.

By incorporating both quantitative assessments, such as survey analysis, and qualitative insights drawn from literature reviews and case study comparisons, the research ensures a balanced and in-depth approach. The structured surveys provide firsthand data on pilgrim behavior, concerns, and preferences, while secondary sources contribute broader insights into globally recognized best practices. The integration of these findings enables the study to formulate well-rounded recommendations tailored to the specific challenges of Char Dham.

6.6   Primary Data Collection

Primary data collection involved conducting structured surveys at multiple locations across the Char Dham circuit, including temple premises, waiting areas, accommodation sites, and transit hubs such as bus stations in Haridwar and Rishikesh. The surveys aimed to capture a diverse range of perspectives by assessing the demographic composition of pilgrims, their travel experiences, the difficulties they encountered due to overcrowding, their awareness of sustainability initiatives, and their willingness to support regulatory measures for more efficient pilgrimage management.

6.7   Survey Structure

The survey was meticulously designed to cover different aspects of the pilgrimage experience, allowing for an in-depth assessment of key issues. The demographic section gathered information about age, gender, occupation, and place of residence, ensuring representation from diverse backgrounds. Questions related to pilgrimage experience focused on travel frequency, preferred transport modes, and duration of stay, providing insights into visitor patterns.

6.8   Sample Size and Participant Breakdown

A total of 110 survey responses were collected, ensuring a well-distributed sample across various demographic categories. The respondents were carefully selected to include individuals who had visited at least one of the four Char Dham sites – Badrinath, Kedarnath, Gangotri, and Yamunotri.

The results from this primary data collection reinforce the need for structured visitor management, improved infrastructure, and enhanced sustainability measures to ensure a balanced approach to pilgrimage tourism in the region.

6.9   Secondary Data Collection

Secondary data was gathered through an extensive literature review focusing on sustainable pilgrimage tourism and visitor management strategies at major religious sites worldwide. Academic publications, government reports, and policy documents were analyzed to assess the effectiveness of various sustainability interventions in pilgrimage tourism.

6.10   Literature Review and Comparative Analysis

The literature review examined studies on environmental degradation linked to religious tourism, particularly in sensitive ecological zones like the Himalayas. Additionally, research on overtourism in Indian pilgrimage sites, such as Varanasi, Tirupati, and Mathura, was analyzed to understand the broader implications of unregulated visitor influx on local communities and temple operations. Furthermore, case studies on Mecca, Vatican City, and Camino

de Santiago were reviewed to assess how these sites have successfully implemented sustainable tourism practices, including digital entry systems, controlled visitor access, and eco-tourism initiatives.

6.11   Global Pilgrimage Site Comparisons

A comparative analysis of Char Dham with other globally significant pilgrimage destinations provided crucial insights into effective pilgrimage management strategies. For instance, the quota system and controlled entry permits implemented during Hajj in Mecca offer a structured approach to regulating pilgrim numbers, which could be adapted for Char Dham during peak seasons. Similarly, the sustainable trail infrastructure and community-driven tourism model of Camino de Santiago highlight the potential benefits of engaging local communities in tourism management. Vatican City's digital ticketing and time-slot entry system demonstrate how technology can be leveraged to minimize congestion at religious sites.

By examining these case studies, the research identifies practical interventions that can be tailored to the unique geographical and cultural context of Char Dham, providing a foundation for sustainable pilgrimage tourism management.

6.12   Data Integration and Analysis

To synthesize the primary and secondary data findings, statistical analysis was applied to survey responses to identify trends and correlations between visitor experiences and sustainability challenges. Additionally, the comparative study provided insights into practical interventions that could be tailored to the unique geographical and cultural landscape of Char Dham.

By integrating survey-based evidence with international best practices, the research establishes a strong foundation for developing sustainable policies to enhance pilgrimage management while preserving the sanctity of Char Dham.

6.13   Results and Analysis

This study, based on extensive survey responses and field observations, presents a well-rounded view of the challenges posed by overcrowding at the Char Dham pilgrimage sites. By examining pilgrim behavior, infrastructure capacity, and sustainability efforts, the findings offer insights into balancing religious tourism with environmental and cultural preservation.

One of the key takeaways is how overcrowding affects the spiritual and logistical aspects of the pilgrimage. Survey data reveals that long waiting times, congestion at temple entry points, and limited access to rituals significantly impact pilgrim satisfaction. While devotees accept certain inconveniences as part of their spiritual journey, many express concerns about declining comfort, restricted temple access, and inadequate facilities. Understanding these behavioral patterns helps pinpoint areas where improvements can enhance both convenience and spiritual fulfillment.

The study also assesses the effectiveness of sustainability measures at Char Dham. Waste management initiatives, crowd control strategies, and conservation policies have been introduced, but their implementation remains inconsistent. Field observations highlight gaps such as ineffective waste disposal, insufficient public amenities, and environmental strain due to infrastructure expansion. By comparing Char Dham with similar religious destinations, the research identifies best practices that could be adapted to this unique Himalayan setting.

Perhaps the most pressing issue is finding feasible solutions to regulate visitor numbers without diminishing the pilgrimage's spiritual essence. The study evaluates measures like controlled entry permits, timed visitation slots, and infrastructural upgrades, considering their practicality in Char Dham's religious and geographic context. While global examples provide valuable lessons, any intervention must align with the pilgrimage's inclusive and sacred nature.

Overall, these findings offer a holistic understanding of how overcrowding shapes the Char Dham experience and propose well-researched strategies for sustainable management. The insights presented in this study serve as a practical resource for policymakers, religious institutions, and local stakeholders, guiding them in implementing effective solutions that uphold both the sanctity and longevity of the Char Dham Yatra.

6.14   Pilgrim Behaviour and Satisfaction

The survey data provides valuable insights into the demographics, motivations, and challenges faced by pilgrims at Char Dham. A significant portion of visitors falls within the 26–50 age group, with an almost equal representation of men and women, highlighting the widespread appeal of the pilgrimage. While religious devotion remains the primary motivation, an increasing number of pilgrims are also drawn by cultural exploration and adventure tourism. Both first-time and repeat visitors contribute to the growing footfall, with 40% of respondents having undertaken the journey multiple times. Overcrowding emerged as a pressing issue, with 85% of pilgrims experiencing congestion, particularly at Kedarnath and Badrinath, leading to long waiting times, accommodation shortages, and traffic delays, especially during peak festival months. Many travelers, particularly the elderly and those with mobility challenges, struggled with inadequate infrastructure, poorly maintained pathways, and a lack of rest areas, which significantly impacted their experience. Despite these difficulties, most respondents still found deep spiritual fulfillment but expressed a strong need for better infrastructure and management. These findings emphasize the urgency of implementing effective visitor regulation, enhancing accessibility, and developing sustainable infrastructure to preserve the sanctity of Char Dham while accommodating the growing number of pilgrims.

6.15   Effectiveness of Existing Sustainability Measures

While several sustainability initiatives have been introduced at Char Dham, their impact remains limited due to inconsistent enforcement and a lack of awareness among pilgrims, with only 40% familiar with existing waste management programs. Efforts such as plastic bans and eco-friendly transport have struggled to achieve meaningful success, primarily due to weak implementation and communication gaps. Comparisons with global pilgrimage destinations offer valuable lessons – Mecca's structured crowd management and Camino de Santiago's community-driven conservation efforts have effectively reduced congestion and minimized environmental impact. Char Dham could benefit from adopting similar approaches by integrating visitor scheduling, strengthening waste management practices, and increasing local community involvement. Strengthening enforcement measures and conducting widespread awareness campaigns would play a crucial role in preserving the spiritual and environmental integrity of the pilgrimage while ensuring its long-term sustainability.

6.16   Feasibility of Sustainable Solutions

The study's findings reveal that most pilgrims are open to the implementation of sustainable practices, as long as these measures do not interfere with religious traditions. More than 60% of respondents expressed support for the introduction of regulated visitor slots, believing that scheduled temple visits could serve as an effective means of managing congestion. This sentiment reflects a growing recognition among pilgrims that structured entry systems could enhance their spiritual experience by reducing overcrowding and long waiting times. Additionally, 70% of respondents emphasized the need for improved public transport, acknowledging that better transit options could significantly ease vehicular congestion on the narrow roads leading to the Char Dham sites.

Economic considerations also emerged as a crucial factor in assessing the feasibility of sustainable tourism solutions. Many local businesses rely heavily on the peak pilgrimage season for their livelihood, making it challenging to implement measures that might inadvertently reduce visitor numbers. As a result, any proposed interventions must carefully balance economic stability with conservation efforts to

ensure long-term sustainability. Strategies such as promoting off-season travel and encouraging homestays as alternative accommodations were identified as viable solutions for distributing tourism activity more evenly throughout the year. By incentivizing visitation during less crowded periods and integrating community-based tourism initiatives, these approaches could help mitigate the adverse effects of mass tourism while fostering economic resilience in the region.

Overall, the study underscores the urgent need for structured and well-coordinated interventions to address the mounting pressures associated with large-scale pilgrimage tourism. By examining pilgrim behavior, evaluating the effectiveness of existing sustainability measures, and assessing the practicality of proposed solutions, this research lays a strong foundation for developing targeted recommendations. These insights can guide policymakers, religious authorities, and local stakeholders in implementing strategies that safeguard the sanctity of the Char Dham pilgrimage while ensuring its long-term viability for future generations.

6.17   Dynamics of Overcrowding

The challenge of overcrowding at the Char Dham pilgrimage sites has intensified over the years, largely due to increased accessibility, the expansion of digital tourism promotions, and evolving socioeconomic dynamics. Official records indicate that the number of pilgrims undertaking the sacred journey has grown exponentially, rising from approximately 200,000 in the 1980s to over 3 million annually. This surge in visitors has placed an immense burden on the existing infrastructure while also exerting significant pressure on the region's natural resources.

Several interrelated factors contribute to this growing congestion. The development of improved road networks, the expansion of helicopter services, and the widespread availability of digital pilgrimage guides have made the Char Dham Yatra more accessible to a broader demographic. Additionally, the influence of social media and large-scale religious campaigns has further amplified interest in pilgrimage tourism, inspiring a greater number of individuals to embark on the journey.

The issue of overcrowding is most pronounced during peak pilgrimage months, particularly between May and October. Religious festivals such as Akshaya Tritiya and Janmashtami see a sudden and sharp influx of devotees, leading to thousands of visitors arriving at the sites simultaneously. Spatial analysis indicates that congestion tends to be particularly severe in certain high-traffic areas, including temple entrances, accommodation centers, and transport hubs. The narrow pathways leading to the shrines further exacerbate the situation, as large crowds often struggle to navigate the steep and rocky terrain.

Beyond logistical difficulties, the overwhelming number of visitors significantly impacts the spiritual experience of pilgrims. Many devotees express concern that the sheer volume of people makes it challenging to engage in moments of quiet reflection and prayer. The presence of noise pollution, stemming from temple announcements, loudspeakers, and commercial vendors, further detracts from the intended atmosphere of peace and devotion, thereby diminishing the deeply personal and sacred nature of the pilgrimage.

Despite these challenges, efforts to regulate visitor numbers have frequently been met with resistance. Many pilgrims feel that imposing restrictions would contradict the open and inclusive essence of religious pilgrimage. However, similar concerns were raised at other globally significant religious destinations before structured visitor management systems were successfully introduced. For instance, the Vatican has implemented controlled entry timings to prevent overcrowding at St. Peter's Basilica, while the Hajj pilgrimage follows a well-regulated quota system to manage the immense annual influx of worshippers. These examples illustrate that carefully designed and sensitively implemented visitor management strategies can help safeguard the sanctity of religious experiences while ensuring a more structured and fulfilling pilgrimage journey.

Addressing these complex challenges necessitates a balanced approach – one that respects religious traditions while incorporating practical measures to enhance visitor experiences and mitigate overcrowding. Pilgrim education

initiatives, improved scheduling systems, and strategic investments in infrastructure could play a pivotal role in addressing these concerns. By adopting a thoughtful and well-coordinated approach, it is possible to preserve both the spiritual significance of the Char Dham Yatra and the long-term sustainability of these revered pilgrimage sites for future generations.

6.18 Environmental Impact

The Char Dham region, nestled within the fragile ecosystem of the Himalayas, is facing mounting environmental challenges due to the uncontrolled rise in pilgrim footfall. The increasing number of visitors, coupled with inadequate waste management and unsustainable tourism practices, has placed immense pressure on the area's natural resources and disrupted its ecological balance. As pilgrimage numbers continue to rise, the strain on local biodiversity, water bodies, and air quality has become more severe, raising significant concerns about the long-term sustainability of this sacred landscape.

One of the most pressing environmental concerns is waste accumulation. The sheer volume of solid waste generated by pilgrims – including plastic bottles, food wrappers, and disposable religious offerings – has overwhelmed existing waste management systems. Although efforts have been made to ban single-use plastics at pilgrimage sites, enforcement remains inconsistent, and the available disposal facilities are insufficient to handle the daily influx of litter. Surveys indicate that nearly 70% of pilgrims are unaware of proper waste disposal guidelines, leading to widespread dumping along trekking routes, temple premises, and riverbanks. This uncontrolled waste not only detracts from the natural beauty of the region but also poses serious environmental hazards, threatening wildlife and contaminating vital water sources.

Water pollution is another critical issue affecting the sacred rivers of the Char Dham circuit. The Yamuna and Ganga, which hold immense religious significance, frequently suffer from contamination due to ritual offerings, untreated sewage, and runoff from makeshift settlements. Many devotees immerse flowers, incense, and other offerings – some of which contain non-biodegradable materials such as plastic and synthetic fabrics – directly into the rivers. This practice, combined with inadequate waste filtration systems, has contributed to increased water toxicity. Studies show that water quality in these rivers deteriorates significantly during peak pilgrimage months, heightening health risks for both pilgrims and local communities.

Deforestation also poses a serious environmental threat, driven by the need to accommodate growing visitor numbers. Large-scale forest clearing for the construction of roads, lodges, and commercial establishments has led to extensive habitat destruction, putting immense pressure on the region's delicate ecosystems. The loss of tree cover accelerates soil erosion and increases the risk of landslides and flash floods, endangering both human settlements and wildlife. Ecological assessments indicate that such unsustainable infrastructure expansion has had a detrimental impact on species like the Himalayan musk deer and the snow leopard, both of which rely on the region's forested landscapes for survival.

In addition to these issues, air pollution has emerged as an increasing concern in the Char Dham region. The reliance on diesel-powered vehicles, coupled with severe traffic congestion on narrow mountain roads, has led to a noticeable decline in air quality. During peak pilgrimage months, carbon emissions from transport vehicles and temporary cooking setups exceed sustainable levels. This problem is particularly pronounced in high-altitude areas, where pollution disperses more slowly. Reports from both pilgrims and residents highlight an increase in respiratory issues, further underscoring the environmental impact of unregulated tourism activities.

Despite these pressing challenges, the implementation of sustainable tourism practices in the region remains limited. While initiatives such as eco-friendly lodging, green trekking routes, and community-led conservation programs do exist, they have yet to be scaled up to effectively counteract the magnitude of environmental degradation. Comparisons with other global pilgrimage sites highlight valuable lessons for Char Dham. For instance, the Camino de Santiago in Spain and Mount Kailash in Tibet have successfully introduced waste segregation programs and eco-certification for tourism facilities, significantly reducing their environmental footprint. Adopting similar models in Char Dham could help mitigate the negative effects of mass tourism while promoting responsible pilgrimage practices.

Preserving both the sacred and ecological integrity of the Char Dham region requires immediate and well-coordinated interventions. Strengthening waste management systems, enforcing stricter pollution control measures, and promoting eco-friendly transport alternatives are crucial steps in ensuring that religious tourism does not come at the cost of environmental destruction. By integrating sustainable practices into pilgrimage management, it is possible to strike a balance between spiritual fulfillment and ecological preservation, ensuring that these revered sites remain accessible and pristine for future generations.

6.19 Infrastructural Strain

The growing number of pilgrims visiting Char Dham has placed immense pressure on its infrastructure, leading to transport bottlenecks, overcrowded accommodations, inadequate sanitation, and limited medical services. Narrow, landslide-prone roads and heavy reliance on private vehicles create severe congestion, while insufficient public transport options make travel difficult. Accommodation shortages during peak seasons force many pilgrims into overcrowded or makeshift lodgings with poor hygiene, and inadequate sanitation leads to environmental degradation, with untreated waste polluting sacred rivers. The region also lacks well-equipped medical centers, leaving many pilgrims vulnerable to altitude-related illnesses and emergencies. To address these challenges, sustainable infrastructural development is essential, including improved road networks, eco-friendly accommodations, enhanced public transport, better sanitation facilities, and expanded healthcare services. Learning from globally managed pilgrimage sites like the Hajj and Vatican City, Char Dham can benefit from structured crowd control, digital booking systems, and environmental initiatives, ensuring a safer, more comfortable, and spiritually enriching experience for future generations.

6.20 Sociocultural Impacts

The Char Dham pilgrimage holds deep sociocultural significance for both pilgrims and the local communities, offering a mix of opportunities and challenges. While pilgrimage tourism contributes significantly to the regional economy, it also raises concerns about cultural shifts, resource conflicts, and transformations in traditional ways of life. As the number of visitors continues to rise, the delicate balance between economic growth and cultural preservation is being increasingly tested.

One of the most significant benefits of pilgrimage tourism is its role in generating employment and supporting local livelihoods. The influx of pilgrims creates a high demand for services such as lodging, food, transport, and religious artifacts, providing a source of income for thousands of residents. In remote areas with limited employment opportunities, many families rely on seasonal tourism as their primary means of financial stability. Local vendors selling handcrafted souvenirs, religious offerings, and traditional food products experience a surge in sales during peak pilgrimage months, helping to strengthen the regional economy. The hospitality and transport sectors, in particular, see notable growth, with small guesthouses, taxi operators, and local guides benefiting from the steady flow of visitors (Gołębieska et al., 2020).

However, the increasing commercialization of the pilgrimage has brought significant challenges. Many traditional spiritual practices, once deeply embedded in local customs, are evolving to accommodate the expectations of modern tourists. The rising popularity of paid VIP darshans, packaged pilgrimage tours, and large-scale promotional religious events has led to concerns that the spiritual essence of the Char Dham Yatra is being overshadowed by commercial interests. What was once an intimate and deeply personal journey of faith has become, for many, a transactional experience. Devotees often express concerns that the spiritual purity of the pilgrimage is being diluted, with sacred rituals at times feeling like exclusive privileges rather than communal religious experiences (Singh and Gowreesunkar, 2019).

Cultural tensions between pilgrims and local communities have also intensified in recent years. The rapid expansion of tourism has driven the large-scale construction of hotels, restaurants, and commercial establishments, often at the expense of traditional settlements and local landowners. The increasing demand for essential resources such as water, food, and firewood during peak seasons frequently leads to shortages for residents, creating friction between visitors and host communities. Additionally, some tourists, unfamiliar with the religious customs and cultural norms of the region, may inadvertently engage in behavior that is perceived as disrespectful, leading to misunderstandings and social tensions (Talukder et al., 2024).

Despite these challenges, there are opportunities to foster a more balanced and harmonious relationship between pilgrims and local communities. Encouraging responsible tourism practices that respect local traditions and ensure fair economic opportunities can help bridge the growing divide. Community-led tourism initiatives, where residents actively participate in managing tourism activities, can empower them while ensuring that the financial benefits are equitably distributed. Implementing fair pricing policies for goods and services can protect small businesses from exploitative commercial practices. Additionally, raising awareness among pilgrims about cultural etiquette, local traditions, and the environmental sensitivities of the region can promote mutual respect and understanding.

By adopting a thoughtful and sustainable approach, it is possible to preserve the cultural and spiritual essence of the Char Dham Yatra while ensuring that local communities continue to thrive. Striking a balance between economic development and cultural preservation will be essential in maintaining harmony between the sacred pilgrimage and the people who call this region home.

6.21 Recommendations

Ensuring the long-term sustainability of the Char Dham pilgrimage requires a well-rounded strategy that tackles overcrowding, environmental impact, and infrastructure limitations. By focusing on visitor management, infrastructural improvements, environmental conservation, community involvement, and technological solutions, the pilgrimage experience can be enhanced while preserving its sacred essence.

One of the most effective ways to manage rising footfall is through structured visitor regulation. Introducing a digital booking system that requires pilgrims to pre-register and secure visit dates can help control daily numbers, preventing overcrowding and optimizing resource use. Additionally, staggered time slots for temple visits – such as morning, midday, evening, and night – could distribute crowds more evenly, minimizing congestion and long queues. Encouraging off-season visits through discounted travel packages and special religious events can further ease pressure during peak months, making the pilgrimage more manageable and comfortable.

Infrastructure improvements are equally crucial. Existing road networks, accommodations, and sanitation facilities often struggle to support the growing influx of pilgrims. Expanding and upgrading roads, developing alternative routes, and improving traffic management can significantly reduce travel delays. In areas prone to bottlenecks, widening roads and constructing bypasses would improve accessibility. To curb congestion, eco-friendly transport options such as electric shuttle buses and ropeways should be expanded, following the successful model implemented at Kedarnath.

Accommodation shortages, especially during peak seasons, can be addressed by expanding state-run pilgrim lodges and incentivizing private investment in sustainable lodging. Encouraging homestays with local families not only provides affordable options but also fosters cultural exchange and supports local economies. Meanwhile, better sanitation and waste management are essential. Increasing public restrooms, setting up biodegradable waste treatment plants, and enforcing strict hygiene standards can create a cleaner, healthier pilgrimage environment.

Given the fragile Himalayan ecosystem, stricter environmental policies are needed. While a plastic ban exists, its enforcement remains weak. Implementing a penalty-based

monitoring system and promoting biodegradable alternatives can significantly reduce pollution. Pilgrim awareness campaigns at key transit points like Haridwar and Rishikesh can educate visitors about eco-friendly practices, ensuring their journey is respectful of the environment. Large-scale tree plantation drives, involving local communities and volunteers, should be prioritized to counter deforestation caused by infrastructure expansion.

Promoting sustainable ritual practices is another way to minimize environmental impact. Pilgrims should be encouraged to use organic offerings instead of plastic-wrapped prasad, which often ends up polluting temple surroundings and rivers. Setting up bio-composting facilities at temple sites can efficiently manage organic waste. With thoughtful environmental initiatives, the Char Dham pilgrimage can thrive without causing lasting damage to its natural surroundings.

Local communities play a vital role in ensuring the sustainability of Char Dham. Their active participation in tourism management ensures that conservation efforts align with local needs while providing economic benefits. Training and certifying local guides can enhance pilgrim experiences by offering cultural insights while also promoting responsible tourism practices. Strengthening community-run enterprises – such as eco-friendly lodges, organic food stalls, and handicraft markets – can create a more balanced tourism model, ensuring that economic benefits reach local families rather than large commercial operators. Investments in healthcare, education, and transport for these communities will further strengthen their role in sustainable tourism.

Advancements in technology offer practical solutions for improving pilgrimage management. A dedicated mobile app providing real-time updates on crowd levels, traffic conditions, weather alerts, and digital queue systems could greatly enhance coordination and convenience for visitors. Geo-tracking can be used to monitor trekkers in remote areas, ensuring their safety, while drone surveillance can help manage crowd movements more efficiently. Smart solutions like these, if integrated thoughtfully, can modernize pilgrimage management without compromising its traditional essence.

By adopting a holistic and sustainable approach, the Char Dham pilgrimage can be preserved for future generations while maintaining its cultural and spiritual significance. Through better planning, responsible tourism policies, and technology-driven innovations, the pilgrimage experience can remain meaningful and fulfilling – without overburdening the environment or local communities.

6.22 Conclusion

The Char Dham pilgrimage holds profound spiritual significance for millions of devotees who embark on this sacred journey each year. However, the increasing number of pilgrims has brought forth a range of challenges, including severe overcrowding, environmental degradation, immense pressure on infrastructure, and sociocultural strains. If these issues are not addressed in a structured and sustainable manner, they could pose a serious threat to the long-term viability of the pilgrimage and the delicate ecological balance of the Himalayan region.

The findings of this study highlight the significant impact of overcrowding on both the pilgrimage experience and the surrounding environment. Many pilgrims endure long waiting times at temples, prolonged traffic congestion along the routes, and a shortage of adequate accommodations, all of which diminish the spiritual essence of their journey. The sheer volume of visitors also places enormous strain on essential services, making it difficult for local authorities to provide sufficient resources, safety measures, and emergency response mechanisms. Consequently, instead of experiencing a deeply fulfilling spiritual journey, many pilgrims face logistical hardships that detract from the sanctity and serenity of the Char Dham Yatra.

Beyond the immediate challenges of overcrowding, the pilgrimage sites and their surrounding landscapes are under immense environmental stress. The unregulated disposal of plastic waste, deforestation caused by infrastructure expansion, and pollution of sacred rivers present serious threats to the ecological health of the region. These environmental concerns not only affect the natural beauty of the pilgrimage sites but also disrupt local biodiversity, impacting the livelihoods of communities

that depend on these ecosystems. If these issues remain unaddressed, the sustainability of the pilgrimage and the well-being of future generations of devotees may be compromised.

A comparative analysis of Char Dham with other renowned pilgrimage destinations around the world offers valuable insights into effective pilgrimage management. Structured policies such as visitor caps, dedicated conservation initiatives, and technology-driven solutions have been successfully implemented in places like Mecca, Vatican City, and the Camino de Santiago. These examples highlight the importance of integrating modern governance tools with traditional religious practices to create a more sustainable and well-managed pilgrimage experience. A key lesson from these global case studies is that striking a balance between religious devotion and responsible tourism management is essential for preserving both the spiritual and environmental sanctity of pilgrimage sites.

Building on these insights, this research puts forth a series of well-defined recommendations aimed at addressing the core challenges affecting Char Dham. One of the most crucial measures is the introduction of visitor management systems that regulate the number of pilgrims allowed at each site per day. Implementing an online booking system, staggering temple entry times, and encouraging off-season visits through incentives can effectively distribute visitor flow and alleviate peak-season congestion. Additionally, significant investment in infrastructure is necessary to enhance road networks, expand eco-friendly transport options, and improve accommodations. Introducing ropeways and electric shuttle services, similar to those in other ecologically sensitive tourist destinations, can reduce vehicular pollution and provide a more sustainable means of transport.

A key priority for ensuring the long-term sustainability of the Char Dham pilgrimage is strengthening environmental conservation efforts. The enforcement of existing plastic bans, the establishment of biodegradable waste treatment plants, and large-scale reforestation initiatives will play a crucial role in preserving the region's ecological integrity. Pilgrim awareness campaigns, conducted at major transit hubs such as Haridwar and Rishikesh, can help educate visitors about responsible tourism practices. Encouraging the use of organic offerings instead of non-biodegradable materials and promoting eco-friendly rituals can further minimize environmental impact.

Community involvement is another fundamental aspect of sustainable pilgrimage management. The active participation of residents in tourism-related activities ensures that economic benefits are distributed equitably while fostering a sense of stewardship over the region's cultural and natural heritage. Providing training programs for local guides, supporting small-scale businesses, and investing in community infrastructure such as healthcare and education can help integrate local voices into the pilgrimage ecosystem in a meaningful way. Empowering local communities will not only enhance the quality of tourism services but also create a more inclusive and sustainable model of pilgrimage tourism.

The incorporation of modern technology into pilgrimage governance presents an opportunity to improve efficiency and the overall visitor experience. Developing a comprehensive Char Dham mobile application with real-time updates on crowd levels, weather conditions, and queue management can significantly enhance visitor planning and logistics. Geotracking systems can be used to monitor and ensure the safety of pilgrims along trekking routes, while drone surveillance can assist in managing crowd movements and traffic congestion more effectively. By adopting these technological advancements, the Char Dham pilgrimage can evolve into a well-regulated and visitor-friendly spiritual journey while minimizing logistical challenges.

As the Char Dham pilgrimage continues to draw millions of devotees each year, its future sustainability depends on the collective efforts of multiple stakeholders. Government authorities, religious institutions, environmental organizations, and local communities must collaborate to implement policies that safeguard the pilgrimage experience without compromising the region's natural and cultural heritage. A holistic approach that integrates tradition with modern sustainability practices will ensure that Char Dham continues to serve as a spiritual sanctuary for generations to come while maintaining its deep connection with nature.

By embracing structured management strategies, investing in sustainable infrastructure, and fostering environmental and community stewardship, Char Dham has the potential to become a global model for responsible pilgrimage tourism. Ensuring the long-term preservation of these sacred sites will not only protect their historical and religious significance but also allow future pilgrims to experience their divine essence in a manner that is both enriching and sustainable.

References

Arora, S., Pargaien, S., Khan, F., Gambhir, A., Khati, K. *et al.* (2023) Management of overcrowding of tourists at 'Kainchi Dham Ashram' in Uttarakhand through sensor technology. In: *Paper presented at 1st International Conference on Circuits, Power, and Intelligent Systems*. DOI: 10.1109/ccpis59145.2023.10291778.

Chandan, S., Pipralia, S. and Kumar, A. (2023) The challenges of urban conservation in the historic city of Puri. *Journal of Urban Regeneration and Renewal* 17(1), 105–121. DOI: 10.69554/bmex6162.

Gołębieska, K., Ostrowska-Tryzno, A. and Pawlikowska-Piechotka, A. (2020) Mobility and sustainable religious tourism – accessibility of holy sites. *Sport i Turystyka Środkowoeuropejskie Czasopismo Naukowe* 3(4), 99–117. DOI: 10.16926/sit.2020.03.31.

Khalilov, K.K. (2023) A 'smart' tourist area (destination) model of forming a cluster of pilgrimage sites. *American Journal of Management and Economics Innovations* 5(9), 12–20. DOI: 10.37547/tajmei/volume05issue09-03.

Kowalik, K., Bartoszewski, J., Laniecka, J., Mulia, P., Papso, P. *et al.* (2022) Pilgrimages as a factor of sustainable development. *Journal Pedagogical Almanac* 30(2), 216–225. DOI: 10.54664/nacg7068.

Mohanty, P.P. and Mishra, N. (2021) Overtourism in religious places: Is it a myth or a journey towards faith, a reflection from Golden Triangle (Bhubaneswar-Puri-Konark) of Odisha, India. In: Sharma, A.

and Hassan, A. (eds) *Overtourism as Destination Risk: Impacts and Solutions*. Emerald Publishing, Bingley, UK, pp. 235–260. DOI: 10.1108/978-1-83909-706-520211016.

Saeedi, A. (2022) Community participation in conservation proposals of Islamic pilgrimage sites. In: *Paper presented at the 38th Annual Conference of the Society of Architectural Historians Australia and New Zealand*. DOI: 10.55939/a4025pfdgv.

Sharma, R. (2020) Study of impact of overtourism on local society at pilgrimage destinations with reference to Mathura and Vrindavan. *International Journal of Modern Agriculture* 9(3), 410–418. DOI: 10.17762/ijma.v9i3.165.

Shashwat, K., Shashwat, A. and Yadav, V. (2024) Sustainable technological innovations in festival tourism: A case study from Uttarakhand. In: Kumar, P., Gupta, S.K., Korstanje, M.E. and Rout, P.C. (eds) *Managing Tourism and Hospitality Sectors for Sustainable Global Transformation*. IGI Global, Hershey, PA, pp. 195–211. DOI: 10.4018/979-8-3693-6260-0.ch014.

Singh, R.P.B. and Rana, P.S. (2023) *Geography of Hindu pilgrimage places (Tīrthas) in India. In: Geography of World Pilgrimages: Cultural, Social and Territorial Perspectives*. Springer Nature, Cham, Switzerland, pp. 279–322. DOI: 10.1007/978-3-031-32209-9_14.

Singh, T.V. and Gowreesunkar, G. (2019) Transformation of Himalayan pilgrimage: A sustainable travel on the wane. *Journal on Tourism and Sustainability* 2(2), 37–45.

Talukder, M.B., Kumar, S., Kaiser, F. and Mia, M.N. (2024) Pilgrimage creative tourism: A gateway to sustainable development goals in Bangladesh. In: Hamdan, M., Anshari, M., Ahmad, N. and Ali, E. (eds) *Global Trends in Governance and Policy Paradigms*. IGI Global, Hershey, PA, pp. 285–300. DOI: 10.4018/979-8-3693-1742-6.ch016.

Yadav, C.S. (2024) Assessing the sustainability of pilgrimage destinations using GSTC criteria: A case study of Varanasi, Uttar Pradesh. In: Pitchaimani, M., Leelawati, K. and Das, S.K. (eds) *Futuristic Trends in Management*, Vol. 3, Book 26. Iterative International Publishers, Karnataka, India, pp. 118–134. DOI: 10.58532/v3bhma26p2ch4.

Further Reading

Catherine, S., Joyce, S., and Rani, M. (2023) Rethinking sustainable-oriented pilgrimage tourism in Tamil Nadu. In: Gupta, S.K., Aragon, L.C., Kumar, P., Madhurima, S., and Ramasamy, R. (eds) *Management and Practices of Pilgrimage Tourism and Hospitality*. IGI Global Scientific Publishing, Hershey, PA, pp. 92–98. DOI: 10.4018/979-8-3693-1414-2.ch007.

Chandan, S. and Kumar, A. (2019) Challenges for urban conservation of core area in pilgrim cities of India. *Journal of Urban Management* 8(3), 472–484. DOI: 10.1016/j.jum.2019.05.001.

Kavitha, V.S. and Firoz, M. (2022) Benchmarking sustainability of pilgrimage cities: A case of three cities in Tamil Nadu, India. *Benchmarking* 30(9), 2967–2992. DOI: 10.1108/bij-09-2021-0547.

Shinde, K.A. and Olsen, D.H. (2023) Reframing the intersections of pilgrimage, religious tourism, and sustainability. *Sustainability* 15(1), 461. DOI: 10.3390/su15010461.

Stobdan, J. (2023) Pilgrimage tourism as a means to attain sustainable development. In: Gupta, S.K., Aragon, L.C., Kumar, P., and Ramasamy, R. (eds) *Prospects and Challenges of Global Pilgrimage Tourism and Hospitality*. IGI Global, Hershey, PA, pp. 213–222. DOI: 10.4018/978-1-6684-4817-5. ch016.

Sutarya, I.G. and Widana, K.A. (2024) Space of conflict in the reproduction of Hindu sacred places into spiritual destination. *Space and Culture, India* 12(1), 110–120. DOI: 10.20896/saci.v12i1.1493.

7 Community-Based Approaches to Sustainable Municipal Waste Management in Indonesia

Hari Harjanto Setiawan[1], Rizal Akbar Aldyan[1]*, Siti Khoiriyah[2], Chanel Tri Handoko[2], Widodo Aribowo[2] and Agung Hidayat[2]

[1]National Research and Innovation Agency (BRIN), Indonesia; [2]Sebelas Maret University, Indonesia

Abstract

Indonesia's escalating waste crisis, driven by urbanization and population growth, underscores the need for sustainable, community-based waste management (CBWM). Despite policies like Law No. 18/2008 promoting community involvement, gaps remain in participation, institutional coordination, and financial incentives. This chapter analyzes CBWM's effectiveness in addressing these gaps and advancing sustainable waste management. Using qualitative methods, including focus group discussions (FGDs) in three districts, this study identifies key success factors such as leadership, policy support, and circular economy integration. Findings reveal that while waste banks and grassroots initiatives foster engagement, regulatory inconsistencies, and limited technological adoption hinder scalability. Implications highlight the need for strengthened multi-stakeholder governance, enhanced digital tools, and targeted behavioral interventions. Bridging policy and practice through innovative, community-driven solutions is essential for improving waste management efficiency. CBWM presents a transformative approach to achieving long-term environmental resilience in Indonesia.

7.1 Introduction

Rapid urbanization, rapid population growth, and limited infrastructure make urban waste management a significant challenge for a country such as Indonesia (Kristyawan *et al.*, 2021). Cities like Jakarta, Surabaya, and Bandung generate millions of tons of waste yearly, but existing waste management systems are often not equipped to handle the growing volume (Noegroho *et al.*, 2021). This generates a lot of waste discarded into landfills, rivers, or even the ocean, contributing to environmental degradation and increasing risks to human health (Rosesar and Kristanto, 2020).

Moreover, the lack of public awareness of the significance of waste sorting and reduction worsens the issue.

Waste is too complex to lose at the end with conventional approaches that only address final disposal and drive waste from one point to another (Abidin *et al.*, 2021). This system is often subject to technical, financial, and institutional constraints. Poor waste management infrastructure and limited recycling facilities are critical barriers to a more sustainable management system (Amasuomo and Baird, 2016). Furthermore, waste management policies are frequently inefficient due to poor government, private sector, and community coordination.

*Corresponding author: riza014@brin.go.id

© CAB International 2025. *Municipal Waste Management: Policies and Strategies*
(eds S. Kumar *et al.*)
DOI: 10.1079/9781836990666.0007

However, a community-based approach has provided a novel and practical alternative to tackle these challenges (Amheka *et al.*, 2015). Community-based waste management is a form of community management that includes the stages of reducing, sorting, and recycling or managing organic waste (Atyadhisti and Sarifudin, 2019). This model accounts for the principle of community empowerment and inclusion, decreasing waste volume and providing economic and social benefits to the local communities (Yousefloo and Babazadeh, 2019).

Several successful community-based initiatives in Indonesia deal with waste. For instance, the waste bank runs in all regions, where communities sort waste for financial rewards (Amheka *et al.*, 2015). Moreover, as a by-product, the treatment of organic waste using compost and biogas plants is spreading among communities to utilize and minimize residual waste to be disposed of, producing a beneficial and valuable commodity for agriculture and renewable energy.

Despite its enormous potential, community-based waste management still encounters many challenges (Towolioe *et al.*, 2020). These programs are needed for mitigation strategies and adaptation (Santoso and Farizal, 2019). However, they face various challenges, such as limited funding, inadequate policy support, and social resistance to behavioral change. Program sustainability is also an issue if the initiative does not possess a strong institutional base and management mechanism.

The Indonesian government has implemented several regulations and programs to promote community-based waste management, including the Clean Indonesia Movement and plastic waste reduction programs. However, a much more integrated and collaborative approach is needed to make these policies effective, starting with better coordination between local governments, the private sector, and civil society, where the last can play a vital role.

Community-based approaches in urban waste management in Indonesia are becoming increasingly important in facing the challenges of rapid population growth and urbanization. However, the effectiveness of this strategy still needs to be thoroughly evaluated to ensure its sustainability. This study examines how community-based approaches can improve sustainable waste management systems and identify key factors that determine their success in various regions of Indonesia. In addition, this study will explore the role of government policies, community participation, and private sector involvement in supporting the implementation and sustainability of these initiatives. The integration of circular economy principles is also a crucial aspect of optimizing community-based waste management, so this study will examine how this concept has been implemented in Indonesia.

Furthermore, with technological innovation and digitalization, this study will evaluate how technology can improve waste management efficiency and strengthen community participation. Although various initiatives have been implemented, funding, regulatory support, and changes in community behavior are still challenges. Therefore, this study will examine the main obstacles in implementing community-based waste management and propose effective strategies to overcome them. With this approach, this study is expected to significantly contribute to strengthening more sustainable waste management policies and practices in Indonesia.

7.2   Theoretical Framework

The increasing waste volume due to urbanization and population growth necessitates community-based waste management (CBWM) in Indonesia. This approach emphasizes community participation, enhances efficiency, and fosters collective ownership of environmental sustainability. Governance plays a crucial role in CBWM success, requiring synergy between national and local authorities to ensure adequate infrastructure and resources (Latanna *et al.*, 2023; Fariz *et al.*, 2024).

A key CBWM model is the waste bank system, which facilitates waste sorting, recycling, and reuse. This is demonstrated in Makassar, where integration with city policies improved waste management efficiency (Kubota *et al.*, 2020). Sukunan, Yogyakarta, also implemented a zero-waste approach, encouraging waste sorting at the source and organic waste composting, aligning with circular economy principles (Kurniawan *et al.*, 2021). These initiatives not only reduce landfill waste but also create economic opportunities.

However, structural challenges persist. Low environmental literacy and weak policy support hinder public participation in sustainable waste management (Luthfiani and Atmanti, 2021). Thus, government–community collaboration is essential for holistic waste management strategies encompassing education, infrastructure, and policy integration (Fariz *et al.*, 2024).

The 3R (reduce, reuse, recycle) strategy further strengthens CBWM by promoting sustainable practices, as seen in Cilandak, Jakarta, where reusable packaging and composting initiatives reduced waste at the source while generating economic benefits (Sabela *et al.*, 2022; Ekananda and Sumendar, 2023).

CBWM presents a viable solution for Indonesia's waste crisis. Its success relies on enhanced public education, supportive policies, and multi-stakeholder collaboration. A comprehensive and coordinated approach can establish CBWM as a critical pillar in achieving sustainable environmental management.

7.3   Method

7.3.1   Research design

This research adopts a qualitative method of focus group discussions (FGDs) to study community-based waste management programs in Indonesia holistically (Fajarwati *et al.*, 2020). Due to its aim of accessing rich stakeholder perspectives, our qualitative methodology enabled detailed accounts of the development, barriers, and emerging success factors in community-led waste management program implementation (Amheka *et al.*, 2015).

7.3.2   Study areas

The FGDs were conducted in three different areas, namely, Eromoko District, Wonogiri Regency; Jaten District, Karanganyar Regency; and Mojolaban District, Sukoharjo Regency, which have successfully implemented community-based waste management programs. These areas were deliberately selected to reflect diverse socio-economic and environmental contexts,

including urban, peri-urban, and rural areas. This approach ensures a holistic understanding of the challenges and opportunities that influence the effectiveness of community-based waste management across different geographic and demographic landscapes (Vergara and Tchobanoglous, 2012). The research was conducted from January to August 2024.

7.3.3   Participant selection

Each FGD consisted of 8–12 participants, strategically selected to ensure a balanced representation of key stakeholders, including:

1. Community leaders: local waste management programs such as waste banks and composting.
2. Policy makers: public officials in local government agencies who implement waste management policies and regulatory frameworks.
3. Practitioners: what our community calls 'waste management workers' such as collectors and recyclers.
4. Local officials: municipal waste management departments are responsible for implementing and regulating waste management.
5. NGOs: nongovernmental organizations that promote environmental sustainability and waste management.

Participants were purposefully sampled to include diverse perspectives with rich experiential insights toward community-based waste management. Ethical considerations, such as informed consent and confidentiality, were followed throughout data collection to maintain the integrity of the research and participant well-being.

7.3.4   Data collection

The focus group discussions (FGDs) adopted a semi-structured approach, which provided sufficient flexibility to allow participants to narrate their life experiences while ensuring that the major thematic areas were systematically covered (Maharani *et al.*, 2019). Discussions were guided by a structured set of questions that explored:

1. Community participation: the levels of engagement, motivating drivers, and challenges that community members face.
2. Financial sustainability: all initiatives are driven by funding mechanisms, cost-recovery strategies, and economic incentives.
3. Policy support: governmental policies, their enforcement, and their gaps.
4. Technological integration: adopting innovative technologies to improve the effectiveness of waste management.
5. Scalability challenges: principal obstacles and prospects for scaling up and replicating successful community-led waste management systems.

Informed consent was obtained from participants, and each focus group discussion session was audio-recorded and lasted between 90 and 120 min. In addition to the recorded data, field notes were used to describe additional nonverbal cues, contextual nuances, and conversation highlights.

7.3.5 Data analysis

Thematic analysis was employed to examine the FGD transcripts systematically. The analytical process involved:

1. Transcription and familiarization: verbatim transcription of discussions, followed by iterative readings to identify emerging patterns and themes.
2. Coding and categorization: systematic coding of recurrent themes was subsequently categorized into broader thematic constructs.
3. Interpretation and synthesis: analytical synthesis of themes to construct a comprehensive narrative on community-based waste management initiatives' effectiveness, challenges, and potential.

7.3.6 Validity and reliability

Triangulation was applied by cross-referencing FGD data with policy documents, field observations, and relevant secondary literature to enhance the credibility and robustness of the findings. Furthermore, member checking was conducted, wherein preliminary findings were shared with selected participants for validation and refinement.

By adopting this rigorous qualitative approach, this study offers a nuanced and contextually grounded understanding of community-based waste management in Indonesia. The findings provide valuable insights for policy makers, urban planners, and environmental practitioners, informing the development of more effective and sustainable waste management policies and practices.

7.4 Result

7.4.1 Case study of community-based waste management

7.4.1.1 Community-based waste management in Eromoko district, Wonogiri Regency

The Eromoko community's waste management is integrated waste management. This is necessary so that every stage, from reduction, sorting, transportation, and processing to final disposal, can be carried out efficiently and sustainably.

Optimal waste management can reduce pollution, increase recycling practices, and support a circular economy. In addition, implementing education and policies such as waste banks and environmentally friendly technologies are important factors to ensure the success of waste management, as shown in (Table 7.1). With good collaboration, a clean and healthy environment can be achieved. The sustainable waste management model combines various strategies and technologies to reduce environmental impacts, maintain resource availability, and improve community welfare.

The community, primarily farmers/planters, has a significant role in managing household waste and recycling practices. A total of 45% of respondents answered that their role is vast in these practices. Household waste that is managed into compost to fertilize plants on agricultural or plantation land triggers members to

Table 7.1. Types of waste found around the Eromoko community. Author's own

No.	Waste type	Examples found
1	Organic waste (easily decomposed)	Food waste (rice, vegetables, fruit)
		Leaves and twigs
		Fruit and vegetable peels
		Livestock manure (cows, chickens, goats)
		Straw and agricultural harvest waste
2	Inorganic waste (difficult to decompose)	Plastic (plastic bags, food wrappers, plastic bottles)
		Glass (drink bottles, broken glass)
		Metal (food cans, nails, wire)
		Paper and cardboard (used newspapers, cardboard)
3	B3 waste (hazardous and toxic materials)	Pesticides and their packaging
		Used batteries
		Damaged neon and electronic lamps
		Expired medicines
4	Waste from livestock and fisheries	Livestock manure
		Leftover animal and fish feed
		Fish bones and scales
5	Waste from household activities and home industries	Waste of used cooking oil
		Leftover cloth from small industries (convection, tailors)
		Tofu and tempeh dregs

contribute to managing household waste and recycling practices.

Other roles that are carried out are reducing plastic waste and household waste: 57% of respondents stated that their role is vast in these activities. This is because the production of household waste can be managed by composting it for plant fertility. Plastic waste can be reduced with natural materials, for example, wrapping food with leaves or *besek* (lunch boxes made of woven bamboo).

Waste recycling is reprocessing waste to be reused or converted into new, valuable products. Waste recycling can reduce environmental pollution and provide economic benefits by creating valuable products from materials previously considered waste. Recycling activities carried out by the community in the Eromoko subdistrict include recycling organic waste by composting with leftover food and leaves and animal feed by utilizing leftover vegetables and food processed into alternative feed for livestock such as chickens and fish and recycling inorganic waste by recycling paper to be reused as packaging materials. Recycling textile waste using leftover fabric into creative products such as bags, wallets, and home decorations. Recycling construction waste by utilizing building debris, for example, used concrete and bricks are crushed to be reused as construction materials. Used wood from old buildings can be reused for furniture or biomass fuel.

In the future, Eromoko's waste management needs to use more advanced technology for better handling. The university and related agencies need technology assistance so that the village community feels technology-based waste management. Waste-to-energy (WtE) technology, such as incineration and anaerobic digestion, offers solutions for waste management and renewable energy generation. This technology reduces waste volume and contributes to urban energy needs.

7.4.1.2   Community-based waste management in Jaten District, Karanganyar Regency

The Aisyiyah Community in Jaten District, Karanganyar, is part of the Muhammadiyah women's organization network that has a strategic role in various social, educational, and environmental activities. One of the initiatives carried out by this community is the development of environmental awareness at the local level through household waste management programs, greening, and circular economy-based skills training. In addition, Aisyiyah is also active in women's economic and social empowerment by integrating Islamic values to impact sustainable development in Karanganyar positively.

Currently, household waste management in Jaten District is still not optimal, so intervention is needed to train homemakers to increase awareness and skills in utilizing waste more productively. Based on the results of coordination with various stakeholders, it was agreed that the training program that will be provided to Aisyiyah members of Jaten District includes using scrap fabric to make accessories in the form of brooches that have economic value as well as training in making compost and liquid organic fertilizer (POC).

This community service activity was conducted at the Ar-Rokhim Musholla, Jaten, Karanganyar, on July 13, 2024, at 08.00–13.00 (WIB). Forty Aisyiyah members attended this event throughout Jaten District, including guests and the implementing team. With the theme 'Empowerment of the Aisyiyah Women's Community in Jaten District in Realizing Family Resilience and Facing Climate Change', this activity aims to provide training in economic-based household waste management. This training emphasizes the importance of the role of women in maintaining family resilience and facing the increasingly unknown impacts of climate change.

The implementation of this activity is divided into two sessions. The first session is training on making brooch accessories from household waste guided by NoerHidayati, an accessory small and medium enterprises (SMEs) actor from Laweyan, Surakarta. The second session is training on making compost and liquid organic fertilizer guided by Sulistyo, a doctoral student in environmental science at Sebelas Maret University. This training is expected to increase women's capacity to manage household waste productively while providing added economic value for their families.

Household organic waste management is an important aspect of efforts to reduce negative environmental impacts and support the sustainable use of resources. Science and technology application significantly develops more efficient and environmentally friendly waste processing methods. Composting and making liquid organic fertilizer are two technologies applied in this program.

Compost is a natural process of decomposing organic materials into humus, rich in nutrients, and can be used as plant fertilizer. This technique converts organic waste, such as food scraps, dry leaves, and plant cuttings, into valuable organic fertilizer. In addition, using household composters equipped with sensors and temperature controllers can accelerate the decomposition process, resulting in more optimal fertilizer.

Liquid organic fertilizer is produced from the fermentation of natural organic materials such as compost, animal waste, and plant waste. This liquid contains essential nutrients that can increase soil fertility and support plant growth naturally. Liquid organic fertilizer not only reduces dependence on chemical fertilizers but also contributes to a more sustainable and environmentally friendly agricultural system.

In addition to organic waste management, this activity utilizes inorganic waste for handicraft production. Used paper, such as cardboard or newspapers, can be made into baskets, bags, or wall decorations with economic value. Used plastic can be processed into jewelry, bags, or other accessories. Meanwhile, scraps of cloth and used clothing can be re-sewn into various creative products such as bags, clothes, or blankets. This initiative reduces the volume of waste and opens business opportunities based on a sustainable circular economy. Applying science and technology to producing handicrafts from household waste has great potential. Increasingly advanced plastic recycling technology can process plastic waste into raw materials for craft products. 3D printing and automatic sewing machines can increase production efficiency, while digital applications and e-commerce platforms allow business actors to reach a broader market. In addition, using sensors to monitor product quality and using renewable energy in the production process can support the loss of this craft business.

7.4.1.3 Community-based waste management in Mojolaban District, Sukoharjo Regency

From the questionnaire results, all participants agreed that waste is a problem that must be resolved immediately in Sukoharjo Regency. As many as 79.3% of participants said that people in the Sukoharjo area are very concerned about sustainable waste management, while 10.3% are at the level of being very concerned and even not caring at all.

The survey results show that most participants and managers of waste banks have implemented composter and eco-enzyme technology to manage household organic waste. Other familiar technologies that they often use are liquid organic fertilizers and bio pores. At the same time, many participants still have not implemented kitchen waste-loading technology or biogas. Some respondents have not processed their household waste until now.

Aerobic composters require sufficient air to compost, so they have holes that allow air to enter them. Drenched conditions are avoided in aerobic composters, so the product is only solid compost. Aerobic composters are simple in design but reliable in performance. They are suitable for beginners because they are free from odor and maggots. The composters introduced here are the aerobic composter (designed by the National Research and Innovation Agency [BRIN] team) and the Takakura Composter (designed by Professor Takuka).

Organic waste processed in an aerobic composter consists of:

- soft food waste, be it vegetables, fruits, or various types of cooking leftovers;
- food waste containing gravy or water needs to be drained first; and
- grass and leaf waste.

Bones, leftover meat, leftover fish, and demanding food waste (maize cobs, snake fruit seeds, mango seeds, durian skins, and so on)

should not be composted in an aerobic composter because it will interfere with the process and cause odor.

The compost produced helps loosen or improve the soil and is used as plant fertilizer. Generally, it is used as fertilizer for plants planted by families, such as flower plants, family medicinal plants, fruit trees, etc.

Biopores are holes in the soil that are formed due to various activities of microorganisms in it, such as worms, plant roots, termites, and other soil fauna. The holes will be filled with air and become a place for water to pass through.

Increasing the number of biopore holes can be done by making vertical holes in the soil. The vertical holes are then filled with organic material, which will later be used as a source of energy for organisms in the soil to form more biopores. The vertical holes and bio pores that are formed can be used as artificial water absorption holes and can also be used to process organic waste into compost. This infiltration hole is called a biopore infiltration hole or LRB.

The types of organic waste that can be processed with bio pore holes are not limited to soft food waste but can also be hard food waste (bones, snake fruit seeds, etc.) and vegetables with soup. Leaf and grass waste can also be put into bio pore holes.

Biopore infiltration holes do not require a large yard area for placement. LRB can be placed in water flow areas, around flower pots, fruit plants, etc.

Eco-enzyme is a natural solution that results from the fermentation of sugar and various fruit and vegetable waste in anaerobic conditions (without oxygen). It has various benefits, such as being a natural cleaner and antibacterial, improving soil fertility and restoring polluted waters.

The technique for making an eco-enzyme solution was developed by Dr Rosukon Poompanvong, founder of the Thai Organic Farming Association, who has been conducting research since the 1980s. Meanwhile, Dr Joean Oon, a Naturopathy researcher from Penang, Malaysia, mostly carries out its socialization and promotion globally. The type of waste that is the raw material for eco-enzyme solution is relatively specific: food waste, namely, fresh vegetable and fruit waste, not food waste that has been cooked and started to rot or fatty food waste.

Eco-enzyme solution is brownish with a fresh sour aroma. The colour of the eco-enzyme solution varies from light brown to dark brown, depending on the type of fruit and vegetable waste and the type of sugar used. The manufacturing time of eco-enzyme is three months in tropical areas and six months in subtropical areas. The ratio of eco-enzyme ingredients is one part sugar, three parts fruit/vegetable waste, and ten parts water. Eco-enzyme solution has various benefits, including natural carbolic acid and cleaner, natural liquid soap, natural air purifier, and natural household cleaner.

7.5   Discussion

7.5.1   Policy implications and recommendations for community waste-based management in Indonesia

This chapter discusses policy implications and strategic recommendations for sustainable development. The main focus includes strengthening community participation through education and capacity building, multi-stakeholder collaboration between government, private sector, and NGOs for more effective solutions, and integration of circular economy to reduce waste and optimize resources. This approach will create more inclusive, collaborative, and sustainable policies.

7.5.1.1   *Strengthening community participation*

The crisis of waste management in Indonesia has reached an alarming proportion due to rapid urbanization and growing population, thus calling for holistic and sustainable action. Active community participation is an essential element in the effectiveness and is the foundation of the waste management system. Policies aimed at reducing this challenge must systematically accommodate public participation to enhance efficiency and sustainability in managing waste practices in various areas. As such, this study compiles different parts of the literature to discuss policy implications and develop strategies that can nurture society's function in Indonesia's waste management system.

Regarding regulation, the Indonesian legal architecture related to waste management, especially Law No. 18 of 2008, underlines the importance of community involvement in waste management practices (Pamuji *et al.*, 2023). However, it is also true that despite the regulation providing a solid legal foundation, the extent of community participation in the implementation of waste management is still quite limited. One of the waste reduction strategies is the waste bank program based on litter in the community, which has carried out community-based initiatives as part of the waste reduction efforts (empirical studies). None the less, this scheme's contribution toward the number of waste being reduced is marginal, only reaching 1.7% in 2017 (Ragiliawati and Qomaruddin, 2020). Hence, waste management policies must focus on improving community-based programs' efficiency through intensive training and sufficient resource availability. This move encourages locals to take ownership and share responsibility for keeping domestic waste at the community level.

Community leadership is one aspect that affects community engagement in waste management. Earlier studies suggest that community leader figures are essential in initiating collective participation in waste management programs (Ragiliawati and Qomaruddin, 2020). Hence, strategies to enhance community involvement should incorporate mechanisms supporting community leaders who can mobilize and inspire communities for sound waste management practices. Leadership training programs that enable leaders with technical and managerial skills to facilitate community-based waste management programs can help realize this objective.

A paradigm shift in community attitudes toward waste management, driven by public education campaigns, constitutes a critical dimension of leadership in environmental sustainability. Empirical studies underscore the effectiveness of tailored, evidence-based educational interventions in enhancing awareness and addressing misconceptions regarding sustainable waste disposal practices, as observed in India (Schippers and Pratiwi, 2017). To achieve this, integrating waste management education into school curricula and facilitating community-based workshops can significantly strengthen public engagement in waste management initiatives. Furthermore, educational programs should emphasize the economic benefits of recycling and composting, as these practices reduce waste volumes and create economic opportunities for local communities (Nohong et al., 2024).

In addition, the implementation of digital technology is also an instrument to increase the level of community participation in waste management. Community participation in more responsive and participatory waste management practices can be encouraged through technology-based innovations, such as mobile applications for reporting waste problems and monitoring waste collection schedules (Suryodiningrat and Ramadhan, 2023). As a result, ensuring that implementations and technological solutions are inclusive and applicable to various fields in society through policies in waste management is a critical point to increase the efficiency of waste management systems at the local level.

To strengthen the community's participation in waste management in Indonesia, a multidimensional approach is needed, regulations strengthened, community leaders empowered, public literacy improved through education programs that can run continuously, and technology integrated into waste management mechanisms. The reduction of waste generation through the sustainability of the community can be achieved by introducing habits from the community to separate and manage waste correctly.

7.5.1.2 Enhancing multi-stakeholder collaboration

Such a multi-stakeholder collaboration is strengthened in Indonesia to develop more effective and sustainable waste management strategies. Since the complexity of the waste management system is also driven by the variety of sources generating waste, it was reported that the major contributors to the waste generation were households (62%), traditional markets (13%), and business centers (7%) (Aprilia et al., 2012; Pamuji et al., 2023). These challenges necessitate a collective governance approach that includes not only local governments but also civil society organizations, the business

sector, and local communities to ensure that policies on waste management are effectively applied and carried out.

Collaborative governance approaches effectively foster active stakeholder involvement and enhance the legitimacy of policies. For example, a case study of the Bone Clean Sampah program in Watampone City shows how collaborative leadership can strengthen coordination and distribution of authority between actors for more inclusive and efficient waste management (Gafur et al., 2023). In Makassar and Bantaeng, waste bank policy implementation including communities shows a correlation that more active participation of citizens can be facilitated with the support of the government, which can generate more positive environmental impacts (Fatmawati et al., 2022).

However, the legal frameworks need to be reformed to improve the mechanisms for multistakeholder collaboration and various initiatives already undertaken. Most existing regulations cannot make room for the development of waste generation and disposal in urban regions, which have continued to grow drastically, with waste generation expected to amount to 70 million tons per year by 2022 (Artha et al., 2023). The local conditions should direct their way of dealing with waste, and the informal sector, like scavengers, is integrated into the formal sector waste management system to improve the effectiveness and sustainability of the waste management system, which is a community-based institutional model proposed (Pamuji et al., 2023).

Other than laws, educating and making people aware is essential to effectively implementing waste management policies. A good communication strategy is needed to cultivate collaboration and responsibility for waste management (Tampubolon and Rahayu, 2019; Wijaya et al., 2022). Community participation in better waste management practices is possible through community-based programs (Tampubolon and Rahayu, 2019; Yacadewa and Musa'ad, 2021), waste sorting training, environmental concerns, etc. In addition, synergies among government, business sector, and civil society through strategic partnerships can maximize resource allocation and foster innovation in waste management (Willmott and Graci, 2016; Ikhwan et al., 2020; Chaerani et al., 2024).

There are five features of collaborative governance: (i) cross-sector, (ii) multi-objective, (iii) human values, (iv) interest conflicts, and (v) the diversity of actors. Where possible and required, a changed model in waste management can lead to multi-stakeholder collaboration in Indonesia. It is anticipated that the synergy generated between these actors will not only come to the response to the challenges presented but also be the nucleus of a more sustainable waste management system in the future.

7.5.1.3 Integrating circular economy principles

Opportunities and complex challenges of implementing circular economy principles in waste management in Indonesia, with increasing waste volumes driven by rapid urbanization and evolving consumption patterns, public and private sectors present collective opportunities for innovation in line with integrative circular economy practices that can support more sustainable future for life cycle waste management. The study identifies policy implications and develops strategic recommendations to enhance the effectiveness of waste management in Indonesia, in particular, a holistic approach combining sound governance, community participation, and advanced technology.

As far as governance is concerned, the quickening of the regulatory system plays a pivotal role in expanding the implementation of circular economic principles in waste management. Governance is needed to drive compliance with relevant policies and enhance stakeholder transparency. Municipal solid waste management systems have proven to be some of the most complex governance systems. According to Kurniawan *et al.* (2021), good governance is essential to fill policy implementation gaps in systems of different scales and sectors, thus promoting accountability and integrity. Moreover, a study assessing solid waste management strategies in Indonesia from 2019 to 2021 showed that reviewing existing policies was crucial in pinpointing improved aspects (Gutama and Iresha, 2023). Flexible governance allows room for community-based initiatives such as waste banks, which can bolster community empowerment and establish a recycling culture and waste reduction (Kubota *et al.*, 2020).

While governance is important, community engagement and participation are also at the heart of the transition to a circular economy. Programs that actively involve communities in waste management increase public participation and strengthen the effectiveness of waste reduction efforts. The 3R concept (reduce, reuse, recycle), with the Rukun Warga-based waste bank model, has been successfully utilized by mobilizing the community's resources for a sustainable waste management model (Towolioe *et al.*, 2016). Meanwhile, the NEKAT SA program in Makassar brought out the best of community-based approaches, producing innovative solutions that uphold circular economy principles (Chaerani *et al.*, 2024). These types of programs can create a sense of ownership of local waste management systems, reducing waste volumes and impacting broader sustainable practices.

Technology is also a driver for improving the effectiveness of waste management systems in Indonesia. As reviewed by Manik *et al.* (2024), waste management is an area where introducing innovative technology can optimize waste collection and processing, thereby minimizing operational inefficiencies. For example, Internet of Things (IoT) technology may support real-time data monitoring and management, which will help with data-based decision making and better resource allocation. Moreover, Samadikun *et al.* (2023), proved that eco-enzymes as organic waste fermentation products were an innovative solution that supports environmental sustainability while providing significant economic value.

Implementing circular economy principles in Indonesia's waste management system requires a multidimensional approach that covers governance, community engagement, and innovative technology. This means policy makers should not just focus on designing and implementing policies that address near-term issues related to waste generation but should also consider long-term sustainability. Indonesia has that potential, which needs to be unfolded solely by optimizing the role of the community and technological developments, which are exclusively hindered. This can overcome transitions to an efficient, adaptive, and circular waste management system in Indonesia.

7.6 Conclusion

The challenge of municipal waste management in Indonesia continues to intensify due to rapid urbanization and population growth. This study underscores the critical role of community-based approaches in fostering sustainable waste management systems. Findings from various case studies demonstrate that public participation, community leadership, environmental education, and digital innovation are fundamental drivers in enhancing the effectiveness of community-led waste management programs.

Despite national policies such as Law No. 18/2008 promoting community involvement, several challenges persist, including limited public engagement, insufficient institutional support, and inadequate financial incentives. Initiatives such as waste banks and local leadership programs have shown promising potential in reducing waste volumes and generating socio-economic benefits. However, their long-term sustainability requires stronger policy integration and enhanced multi-stakeholder collaboration.

Adopting circular economy principles presents a viable pathway to optimizing waste management by improving resource efficiency and minimizing environmental impact. Technological innovations, including digital applications for waste monitoring, bio-enzyme-based organic waste processing, and community-driven education programs, can significantly enhance waste management systems.

This study recommends evidence-based policy strengthening, capacity-building programs for local communities, and the integration of eco-friendly technologies to establish a more resilient and sustainable waste management framework. By fostering a more integrated and collaborative approach, community-based waste management in Indonesia can be a crucial pillar in achieving a cleaner and healthier environment for future generations.

References

Abidin, A.Z., Bramantyo, H., Baroroh, M.K. and Egiyawati, C. (2021) Circular economy on organic waste management with MASARO technology. In: *IOP Conference Series: Materials Science and Engineering 1143*, p. 012051.

Amasuomo, E. and Baird, J. (2016) Solid waste management trends in Nigeria. *Journal of Management and Sustainability* 6(4), 36–44. DOI: 10.5539/jms.v6n4p35.

Amheka, A., Higano, Y., Mizunoya, T. and Yabar, H. (2015) An overview of current household waste management in Indonesia: Development of a new integrated strategy. *International Journal of Environment and Waste Management* 15(1), 86–98. DOI: 10.1504/ijewm.2015.066953.

Aprilia, A., Tezuka, T. and Spaargare, G. (2012) Household solid waste management in Jakarta, Indonesia: A socio-economic evaluation. In: Rebellon, L.F.M. (ed.) *Waste Management: An Integrated Vision*. InTech, London, pp. 71–100.

Artha, A., Nurasa, H. and Candradewini, C. (2023) Recognizing and detecting the effectiveness of policy implementation waste management in Indonesia. *Public Policy Journal* 9(3), 201. DOI: 10.35308/jpp.v9i3.7656.

Atyadhisti, A. and Sarifudin, S. (2019) Community-based waste management strategy: A note on community empowerment level in supporting waste bank at Semarang City, Indonesia. In: *Proceedings of the International Conference on Maritime and Archipelago*, pp. 1–6 (Vol. 167).

Chaerani, T., Sayaranamual, S., Atmaja, K. and Youwe, S. (2024) AFT Pattimura CSR Pentahelix collaboration in the implementation of waste management center social innovation program: NEKAT SA (Negeri Katong without waste). *Indonesian Journal of Social Technology* 5(8), 3652–3666. DOI: 10.59141/jist.v5i8.1249.

Ekananda, B. and Sumendar, R. (2023) Socialization of domestic waste reduction program utilizing reusable packaging and waste management through composting. *J-Abdi Journal of Community Service* 3(2), 317–322. DOI: 53625/jabdi.v3i2.5805.

Fajarwati, A., Setyaningrum, A., Rachmawati, R. and Prakoso, B.S.E. (2020) Keys of sustainable community-based waste management (lesson learnt from Yogyakarta City). *E3S Web of Conferences* 200, 02018.

Fariz, R., Muis, R., Anggraini, N., Rachman, I. and Matsumoto, T. (2024) Good environmental governance roles in sustainable solid waste management in Indonesia: A review. *Journal of Community Based Environmental Engineering and Management* 8(8), 45–56. DOI: 10.23969/jcbeem.v8i1.12035.

Fatmawati, F., Mustari, N., Haerana, H., Niswaty, R. and Abdillah, A. (2022) Waste bank policy implementation through collaborative approach: Comparative study – Makassar and Bantaeng, Indonesia. *Sustainability* 14(13), 7974. DOI: 10.3390/su14137974.

Gafur, A., Mappasere, F., Parawu, H., Fatmawati, F. and Samad, M. (2023) Collaborative leadership in waste management: A case study of Watampone City, Indonesia. *Public Governance Administration and Finances Law Review* 8(2), 109–124. DOI: 10.53116/pgaflr.7016.

Gutama, H. and Iresha, F. (2023) Evaluation of solid waste management effectiveness in Indonesia from 2019–2021: A geographic information system analysis. In: *IOP Conference Series: Earth and Environmental Science 1263*, p. 012067.

Ikhwan, Z., Harahap, R., Andayani, L. and Mulya, M. (2020) Partnership as an effort to consolidate waste management initiatives in tourism destinations on small islands: The case of Penyengat Island. In: *Proceedings of the 1st International Conference on Environmental Science and Sustainable Development, ICESSD 2019*, pp. 1–10.

Kristyawan, I.P.A., Wiharja, A.S., Hendrayanto, P.A. and Santoso, A.D. (2021) Update on waste reduction performance by waste-to-energy incineration pilot plant PLTSa Bantargebang operations. In: *International Conferences on Agricultural Technology, Engineering, and Environmental Sciences. IOP Conference Series: Earth and Environmental Science*, p. 012059 (Vol. 922).

Kubota, R., Horita, M. and Tasaki, T. (2020) Integration of community-based waste bank programs with the municipal solid-waste-management policy in Makassar, Indonesia. *Journal of Material Cycles and Waste Management* 22(3), 928–937. DOI: 10.1007/s10163-020-00969-9.

Kurniawan, T., Avtar, R., Singh, D., Xue, W., Othman, M. *et al.* (2021) Reforming MSWM in Sukunan (Yogjakarta, Indonesia): A case-study of applying a zero-waste approach based on circular economy paradigm. *Journal of Cleaner Production* 284, 124775. DOI: 10.1016/j.jclepro.2020.124775.

Latanna, M., Gunawan, B., Franco-García, M. and Bressers, H. (2023) Governance assessment of community-based waste reduction program in Makassar. *Sustainability* 15(19), 14371. DOI: 10.3390/su151914371.

Luthfiani, N. and Atmanti, H. (2021) Waste management service in Indonesia based on stochastic frontier analysis. *Trikonomika* 20(2), 54–61. DOI: 10.23969/trikonomika.v20i2.3952.

Maharani, A., Dewilda, Y., Darnas, Y. and Dewata, I. (2019) Community-based solid waste management planning in the administrative village of Surau Gadang, Padang City. *IOP Conference Series: Earth and Environmental Science* 314, 012017.

Manik, S., Berawi, M., Gunawan, G. and Sari, M. (2024) Smart waste management system for smart and sustainable city of Indonesia's new state capital: A literature review. *E3S Web of Conferences* 517, 05021.

Noegroho, N., Tedja, M. and Primadi, R.S. (2021) New traditional market based on waste management using 3R method (study case: Warung Buncit Jakarta). *IOP Conference Series: Earth and Environmental Science* 794, 012203.

Nohong, M., Harsasi, M., Muharja, F., Samir, S. and Otoluwa, N. (2024) Formulate an incentive model to involve communities' industries in coastal waste management, Makassar, Indonesia. *Journal of Infrastructure Policy and Development* 8(3), 3102. DOI: 10.24294/jipd.v8i3.3102.

Pamuji, K., Rosyadi, S. and Nasihuddin, A. (2023) The legal institutional model of community-based waste management to reinforce multi-stakeholder collaboration in Indonesia. *Kasetsart Journal of Social Sciences* 44(1), 73–82. DOI: 10.34044/j.kjss.2023.44.1.08.

Ragiliawati, R. and Qomaruddin, M. (2020) Role of community leaders as motivator in waste-bank management in Magetan Regency, Indonesia. *Journal of Health Promotion and Health Education* 8(2), 219–227. DOI: 10.20473/jpk.v8.i2.2020.219-227.

Rosesar, J.S. and Kristanto, G.A. (2020) Household solid waste composition and characterization in Indonesia Urban Kampong. *IOP Conference Series: Materials Science and Engineering* 909, 012077.

Sabela, I., Adhaeni, W., Azzah, A. and Ngazizah, N. (2022) Increasing environmental conservation and economic value of waste through waste bank management. *Islamic Journal of Integrated Science Education* 1(1), 73–82. DOI: 10.30762/ijise.v1i1.285.

Samadikun, B., Sudarno, S., Pusparizkita, Y., Hardyanti, N., Pratama, F. *et al.* (2023) Organic solid waste management by producing eco-enzymes from fruit skin in Permata Tembalang. *Journal of*

Precipitation Communication Media and Environmental Engineering Development 20(1), 21–30. DOI: 10.14710/presipitasi.v20i1.21–30.

Santoso, A.N. and Farizal, F. (2019) Community participation in household waste management: An exploratory study in Indonesia. *E3S Web of Conferences* 125, 07013.

Schippers, B. and Pratiwi, A. (2017) Unravelling of waste in a touristic area of Pangandaran from neglecting towards embracing informal waste management practices, West Java, Indonesia. *Journal of Humanities* 29(2), 191–197. DOI: 10.22146/jh.v29i2.26053.

Suryodiningrat, S. and Ramadhan, A. (2023) Integrated solid waste management system using distributed system architecture for Indonesia: An IT blueprint. *International Journal on Advanced Science Engineering and Information Technology* 13(3), 1177–1183. DOI: 10.18517/ijaseit.13.3.17307.

Tampubolon, E. and Rahayu, A. (2019) Collaborative waste management between the community and public infrastructure and facilities (PPSU) officers at the sub-district level (case of waste management in Jembatan Lima Sub-District, Tambora District, West Jakarta administrative city). *Inspiration Journal* 10(1), 1–13. DOI: 10.35880/inspirasi.v10i1.57.

Towolioe, S., Permana, A., Aziz, N., Ho, C. and Pampanga, D. (2016) The Rukun Warga-based 3rs and waste bank as sustainable solid waste management strategy. *Planning Malaysia* 14(4). DOI: 10.21837/pm.v14i4.157.

Towolioe, S., Permana, A.S. and Kadang, H. (2020) Non-conventional options of managing municipal solid waste towards sustainable solid waste management in Makassar City. *IOP Conference Series: Earth and Environmental Science* 447, 012058.

Vergara, S.E. and Tchobanoglous, G. (2012) Municipal solid waste and the environment: A global perspective. *Annual Review of Environment and Resources* 37(1), 277–310. DOI: 10.1146/annurev-environ-050511-122532.

Wijaya, H., Nurasa, H. and Susanti, E. (2022) Implementation of waste management policy in Cimahi City (case study in Leuwigajah waste service area). *Jane – Journal of Public Administration* 13(2), 341. DOI: 10.24198/jane.v13i2.38130.

Willmott, L. and Graci, S. (2016) Solid waste management in small island destinations: A case study of Gili Trawangan, Indonesia. *Téoros Revue De Recherche En Tourisme*. DOI: 10.7202/1036566ar.

Yacadewa, A. and Musa'ad, M. (2021) Implementation of waste management policy in Sentani District, Jayapura District. *Journal of Public Policy* 4(3), 124–142. DOI: 10.31957/jkp.v4i3.2417.

Yousefloo, A. and Babazadeh, R. (2019) Designing an integrated municipal solid waste management network: A case study. *Journal of Cleaner Production* 244, 118824. DOI: 10.1016/j.jclepro.2019.118824.

Further Reading

Anjani, A., Budiaman, B., and Hidayaht, A. (2024) Environmental care behavior through waste bank (study at Teratai Waste Bank Pondok Pucung Sub-district, Pondok Aren District, South Tangerang City). *Advances in Social Humanities Research* 2(5), 721–730. DOI: 10.46799/adv.v2i5.238.

Budiman, B. and Jaelani, A. (2023) The policy of sustainable waste management towards sustainable development goals. *Journal of Human Rights Culture and Legal System* 3(1), 70–94. DOI: 10.53955/jhcls.v3i1.73.

Faizin, M. (2024) The role of social capital for Bumdes in waste management in Miagan village, Mojoagung District, Jombang Regency. *Airlangga Development Journal* 8(1), 1–7. DOI: 10.20473/adj.v8i1.37989.

Masjhoer, J. and Vitrianto, P. (2024) Community engagement in waste reduction: A critical component for Gunung Sewu Geopark conservation, Yogyakarta, Indonesia. *Environmental & Socio-Economic Studies* 12(2), 1–12. DOI: 10.2478/environ-2024-0008.

Miftahorrozi, M., Khan, S., and Bhatti, M. (2022) Waste bank-socio-economic empowerment nexus in Indonesia: The stance of Maqasid al-Shari'ah. *Journal of Risk and Financial Management* 15(7), 294. DOI: 10.3390/jrfm15070294.

Muljaningsih, S., Indrawati, N., and Asrofi, D. (2022) Waste management policy: A study of Malang waste bank in implementing the green economy concept. *The 4th International Conference on Environment, Sustainability Issues, and Community Development (INCRID). IOP Conference Series Earth and Environmental Science* 1098, 012036.

Pratama, A., Kamarubiani, N., Shantini, Y. and Heryanto, N. (2021) Community empowerment in waste management: A meta synthesis. In: *Proceedings of the First Transnational Webinar on Adult and Continuing Education*, pp. 78–83 (Vol. 548).

Setiyaningrum, I., Wati, A. and Suryati, S. (2022) The existence of waste bank management and the impact on the environment and trends of community consumption (case study of the ngudi resik waste bank krecekan, wironanggan, sukoharjo). *Journal on Biology and Instruction* 2(1), 9–19. DOI: 10.26555/joubins.v2i1.6074.

Suwerda, B., Handoyo, S., and Kurniawan, A. (2018) Determinant factors for managing sustainable waste bank in Bantul urban areas. *Sanitation Environmental Health Journal* 10(1), 37–44. DOI: 10.29238/sanitasi.v10i1.776.

8 Industrial Waste Management and Safety Measures

Prerna Mehta*

GD Rungta College of Science and Technology, India

Abstract

Proper management of industrial waste serves two critical purposes: environmental protection and worker safety together with community health sustainability. As industries expand their production outputs have exponentially risen, leading to an imperative need for comprehensive waste management solutions which effectively adhere to established safety policies. The current chapter examines industrial waste management practices that guide the treatment of hazardous and non-hazardous waste and depict waste reduction methods together with recycling approaches and treatment protocols as well as waste elimination systems.

Waste prevention together with resource recovery stands as the first priority in the waste hierarchy while all-inclusive waste management solutions assist organizations to meet national and international regulations. The chapter documents essential safety procedures that industries must implement for public protection, which include risk assessments and emergency preparedness and proper employee safety education.

Real-world demonstration examples highlight successful industrial waste handling procedures and safety measures which reduce environmental damage while creating safer industrial sites. Creative technological practices and a mindful environmental approach combine to enable efficient waste management operations which reduce health dangers.

This chapter serves as a practical guide for industry stakeholders and policy creators together with environmental managers. The text provides proven solutions and practical directives for managing industrial waste while protecting environmental integrity and workplace security.

8.1 Introduction

Industrial waste management stands today as an essential element for both sustainable development and environmental protection worldwide (Shahbaz *et al.*, 2023). Industrial growth to satisfy escalating demand has triggered substantial increases in waste quantity alongside heavier waste materials requiring strategic waste management systems to limit dangerous ecological impacts and adverse human health outcomes (Singh *et al.*, 2024). Recent advancements coupled with new findings within the field demonstrate the critical importance of industries implementing complete waste management systems focused on waste minimization combined with material recovery and safety guarantees (Czekała *et al.*, 2023).

The circular economy model advances waste reduction through principles of design for reuse along with recycling practices. The contemporary approach leads to reduced environmental impacts combined with optimized resource usage to enable profitable industrial performances while maintaining high

*Corresponding author: prernamehta326@gmail.com

© CAB International 2025. *Municipal Waste Management: Policies and Strategies*
(eds S. Kumar *et al.*)
DOI: 10.1079/9781836990666.0008

environmental responsibilities (Sánchez-García *et al.*, 2024). The growth of waste management efficiency and improved screening technologies through artificial intelligence and machine learning has boosted recycling rates throughout waste sorting facilities (Alsabt *et al.*, 2024).

Experimental evidence now shows that effective waste management frameworks stem from collaborative stakeholder relationships which combine regulatory bodies with industries and communities (Ramos, 2024). Through collaboration multiple parties developed detailed environmentally friendly rules which push business entities to advance past minimum standards toward purposeful waste reduction measures.

The continuous improvement in both regulatory standards and technology coexists with a rising demand to train workers regarding waste management processes alongside safety protocols. Promoting employee safety and an organizational environmental stewardship spirit form the basis for this emphasis on focus.

Ambitious industrial waste management remains crucial to reduce environmental damage while protecting public wellness. The use of creative practices together with modern technologies enabled industries to conduct meaningful progress toward building a sustainable future through proper waste management procedures.

8.1.1 Importance of industrial waste management

Effective industrial waste management surpasses regulation needs by delivering vital protection to environmental quality alongside public health benefits and sustainable industrial practices. Growing industries now face rapidly increasing demands to establish effective waste management strategies. The essential function of managing industrial waste includes protection of public health and regulatory adherence and creates positive effects on corporate impression (Yu *et al.*, 2024). Table 8.1 presents an overview of the essential aspects which demonstrate industrial waste management's vital role.

8.2 Types of Industrial Waste

Industrial waste consists of multiple kinds of production-related waste materials which need specific management solutions to handle disposal effectively. Understanding industrial waste classifications alongside its characteristics serves a dual purpose of regulatory compliance and industrial operation sustainability.

8.2.1 Hazardous waste

Waste falls under hazardous classification when it demonstrates substantial threats both to environmental elements and human welfare. Industrial waste falls under the regulatory umbrella of multiple federal frameworks such as the Resource Conservation and Recovery Act (RCRA) within the United States. The Environmental Protection Agency (EPA) categorizes hazardous waste into several types based on their characteristics: ignitable, corrosive, reactive, and toxic. The US manufacturing sector produced approximately 1.9 million tons of hazardous waste in 2022 with recycling or energy recovery methods making up 65% of that total quantity.

Plasma gasification has become a major innovation in hazardous waste management through its capability to transform organic substances into syngas by employing heat at elevated temperatures. A 2021 study demonstrated that plasma gasification technology yields a 90% decrease in hazardous waste volume along with valuable energy generation. Industrial chemicals are now degraded through bioremediation approaches alongside the deployment of commercial applications for microbial cleanup in areas with industrial spills.

8.2.2 Non-hazardous waste

Non-hazardous waste identifies as all materials that neither threaten human health nor environmental safety. Everyday products like paper together with plastics, food waste and packaging materials are included in this category. The EPA reports that non-hazardous waste generation

Table 8.1. Essential aspects in industrial waste management. Author's own.

Key area	Description	Recent discoveries/data	Reference
Environmental protection	The protection of nature through effective industrial waste management represents a fundamental requirement. When waste disposal follows improper methods it creates significant damage to soil water integrity together with air pollution which produces harmful impacts for both natural ecosystems and biodiversity.	Predictions indicate that waste generation rates will increase globally by 70% throughout the next 50 years if present trends remain unaltered. The progressive increase threatens vital natural habitats as well as ecosystems, thus demanding efficient management.	(Gebrekidan et al., 2024)
Public health	The proper control of industrial waste remains a necessity for human health protection. Exposure to dangerous elements in waste exposure causes various health conditions which range from respiratory diseases to skin infections and serious acute poisonings. The failure to properly control industrial waste has frequently created health emergencies within neighboring areas.	Specific health risks caused by industrial waste exposure have been discovered during health reviews therefore it remains essential to establish strong waste management procedures for prevention purposes.	(Padmanabhan and Barik, 2019)
Regulatory compliance	The rising environmental awareness worldwide leads to stricter regulations about waste management participation. Having proper compliance with these regulations helps industries prevent both penalties and legal consequences that support responsible waste management practices.	Various governments have started implement stringent regulatory measures. Businesses must satisfy their legal obligations through compliance while upholding corporate sustainability principles and shows responsibility.	(Debrah et al., 2021)
Resource recovery and sustainability	Effective waste management requires more than just disposal because it features recycling and material recovery to construct circular economic systems. Modern waste management that minimizes raw material requirements helps to preserve natural resources at a time when human numbers increase while resources decrease.	Industrial applications of recent technological advancements have enabled companies to recover important resources from discarded materials. The shift toward waste material recovery brings resource conservation benefits together with economic improvements through resource recovery processes.	(Mattson et al., 2024)
Corporate responsibility and reputation	The public increasingly holds companies responsible to measure their environmental effect. Enhancing both company reputation and attracting environmentally conscious customers alongside investors becomes possible through proper waste management practices.	Examination shows firms that emphasize sustainable practices achieve higher profitability and gain market superiority. Organizations with strong sustainability profiles can develop loyal customers and investor trust which leads to sustainable long-term success.	(Chen et al., 2024)

in the USA totaled 292.4 million tons during 2020 leading to a recycling figure of 69 million tons. The growing demand for sustainability has motivated businesses to use circular economy practices which transform recyclable waste materials into new materials while minimizing their landfill contributions (Groh *et al.*, 2019).

Current recycling systems have received a boost through AI-driven sorting systems and machine learning which optimizes non-hazardous waste management effectiveness. AI applications within recycling facilities achieve approximately 25% higher sorting precision which results in enhanced material recovery alongside reduced recycled waste contamination.

from 2022 to 2050 with present waste management practices. New waste-to-energy technologies serve as catalysts for altering how we handle solid waste. Modern incineration systems with energy recovery methods transform approximately 90% of solid waste into usable energy which lessens our need for landfills (Tihin *et al.*, 2023).

The conversion of food industry waste into biofuels has received increasing prominence in recent times. Between 2018 and 2021 the International Energy Agency reported a 21% growth in biofuel output from solid organic waste, demonstrating rising demand for sustainable power generated from waste resources (Ahmed *et al.*, 2023).

8.2.3   Special waste

Materials that lack hazardous status must undergo specific handling practices and disposal measures because of their distinctive characteristics, which are sorted into a special waste category. The industrial waste stream consists mainly of three primary components: used oil, batteries, and construction debris. Each year in the United States about 1.3 billion gallons of used oil drain from vehicles, yet 50% of this waste is successfully recycled, as annual reports from the EPA confirm (Kareem *et al.*, 2023).

New technologies in special waste management systems now support more efficient recycling methods. Hydrometallurgical techniques lead the way in extracting valuable materials from electronic waste while international e-waste production surpassed 53.6 million metric tons in 2019 despite managing only a 17.4% recycling rate according to Global e-Waste Monitor data (Ankit *et al.*, 2021).

8.2.5   Liquid waste

Liquid waste platforms stem from industrial manufacturing activities yet companies need specific processing methods to make them suitable for waste disposal. Three main waste types exist under this category: manufacturing wastewater alongside cooling water and industrial chemical leakage. The Centers for Disease Control reported that industrial facilities poured roughly 553 million gallons of wastewater into America's surface waters during the year 2020 (Singh *et al.*, 2023).

Membrane filtration technologies represent essential advancements in liquid waste treatment because they enable vital wastewater recycling. Laboratory tests demonstrate that membranes eliminate pollutants at rates surpassing 99% resulting in wastewater suitable for re-instatement in industrial processes, which presented an opportunity to conserve water supply.

8.2.4   Solid waste

During industrial process production solid materials create waste that becomes discarded as solid waste. Metal shavings alongside plastics glass and organic material represent the classified waste types. According to World Bank projections from 2022 global solid waste is predicted to increase to 3.40 billion tons during the period

8.2.6   Gaseous waste

Industrial waste production in gases is one of the tremendous environmental challenges that contribute approximately 24% of carbon emissions into the atmosphere globally, according to the International Energy Agency in 2021. Cement manufacturing alone accounts for about 8% of these emissions due to energy-consuming processes. To curb this, technologies such as carbon capture and storage (CCS) have begun to play a crucial role in addressing this predicament, one example was Petra Nova in Texas, which captures and stores nearly 1.6 million tons of CO_2 from a coal plant.

In addition, industrial facilities are increasingly adopting catalytic converters and scrubber systems to address various gaseous emissions. A catalytic converter converts organic fumes to less harmful gases, while a scrubber removes undesired gases and particulate material from a gaseous effluent. Nowadays, nanoparticles are used mostly; iron nanoparticles are used for the conversion (Mehta *et al.*, 2024).

In conclusion, a combination of technologies – carbon capture and storage, catalytic converter, and scrubbers – will be requisite to tackle industrial gas waste. Such investments not only lessen the environmental impacts of carbon by aiding the efforts against climate change but also make industry competitive with changing sustainability standards (Fatimah *et al.*, 2024).

8.3   Hazardous Vs Non-Hazardous Waste

Waste divisions fall between hazardous materials that threaten health and the environment alongside non-hazardous materials that do not. Understanding waste classification differences remains essential to establish safe waste management procedures which minimize environmental impacts.

8.3.1   Hazardous waste

Hazardous waste is the one that poses a significant risk to human health and the environment, owing to its toxic, corrosive, flammable, or reactive properties. The US Resource Conservation and Recovery Act describes criteria to classify waste as hazardous; yet it comes mainly from manufacturing industries such as chemical production, agriculture, hospitals, small enterprises, and households. In 2022 total hazardous waste generation was 55 million tons mainly from the manufacturing sector which includes

toxic waste loaded with heavy metals and solvents.

Regulatory focus across the board – a growing concern about electronic waste, batteries, and pharmaceuticals – has gained traction since safer methods of processing and recycling have been formulated. Since then, advanced technologies have also emerged for recovery and consuming energy. One such area is plasma arc technology, which transforms hazardous waste into inert materials. A novel trend that is arising includes bioremediation – the use of genetically modified microorganisms to degrade toxic compounds in the environment. Two other common themes running through these various approaches are sustainability and public health considerations, which will drive risk assessments and guides for safe disposal (Fennelly and Perry, 2018).

8.3.2   Non-hazardous waste

Household waste materials which do not create immediate health threats or environmental harm constitute the non-hazardous category. This category encompasses items such as paper, food scraps, textiles, and yard waste. The definition of organic waste is in transition because selected businesses see its vital role in resource recovery operations.

8.3.2.1   Current statistics

The United States produced roughly 292 million tons of non-hazardous solid waste during 2021 because of growing population numbers and rising consumer patterns. The past several years revealed a 10% improvement in the recycling rate for non-hazardous waste mainly because of intensifying public awareness along with municipal programs supporting sustainability.

8.3.2.2   Innovative management practices

Various innovative technologies and management strategies appear recently in non-hazardous waste treatment practices. The implementation of Internet of Things (IoT) devices into smart waste management systems provides real-time bin-level monitoring that benefits recycling costs and air pollution reduction through optimized collection schedule design. Monitored by IoT devices smart waste management solutions track waste bin sizes to optimize collection routing and slashing operational expenses alongside greenhouse gas (GHG) emission reductions. Anaerobic digestion stands as one of the most efficient waste treatment approaches that transforms organic waste into usable energy and nutrient-rich fertilizer. Food waste diversion from landfills reached 60% in the United States under a 2023 pilot community composting initiative (Bhattacharjya *et al.*, 2021).

8.3.3 Key regulatory changes

New regulations for hazardous and non-hazardous waste have cast a new light on sustainable wastes governance worldwide. The Resource Conservation and Recovery Act (RCRA) was enacted in 1976, establishing a system dubbed 'Cradle to Grave'. The established system mandates further tracking of the hazardous waste from the point of generation up to the time of disposal, so as to control its release into the environment. Among the recent amendments to have reinforced global standards on the transboundary movement of hazardous waste under the Basel Convention are stricter notification and consent procedures, which in turn seek to promote cooperation among countries and stem the flow of illegal dumping-related incidents.

Further, extended producer responsibility (EPR) is changing the design and disposal of non-hazardous waste, rendering businesses responsible for the entire life cycle of a product. Before even recycling by the residents, EPR policies incentivize businesses to develop products that are easier to reuse and recycle, thus reducing the pressure on municipal waste systems. Countries with strong EPR practices, such as Canada and EU member states, have reported dramatic recycling rate increases and reduced reliance on landfills. Such legislation provides scope for sustainable waste management and circular economy today and in the future in Australia and other countries (Kumar *et al.*, 2023).

8.4 Waste Management Hierarchy

The waste management hierarchy operates through a systematic order which evaluates different waste management approaches against environmental effects and sustainability criteria. It typically consists of five key strategies. Five fundamental strategies are prevention followed by minimization then involves reuse along with recycling before disposal action takes place. The following analysis provides in-depth explanations of these strategies along with contemporary industry data.

8.4.1 Waste prevention

Waste prevention marks the strongest strategy by maintaining waste amounts at zero. The goal of product design combined with process evaluation and system construction is to develop methods which create minimal waste while supporting sustainability.

8.4.1.1 Recent developments

Today's organizations put renewable resource utilization together with recyclable material use at the forefront of their sustainable product design initiatives. The electronics sector's initiatives mandate manufacturers to develop repairable and upgradable products which minimizes electronic waste. The recent surging investment wave for circular economy practices during 2023 prioritizes waste prevention as a vital component of economic expansion.

8.4.2 Waste minimization

Manufacturing operations and consumer practices create waste which reduction focuses on both waste amount reduction and decreasing waste toxicological content.

8.4.2.1 Innovative approaches

For several years numerous industries have implemented lean manufacturing approaches which prioritize manufacturing with reduced waste and increased efficiency. Organizations use value stream mapping to detect and analyze waste inside their operational flows. Using reclaimed fabrics combined with innovative production technologies allows the textile industry to achieve waste reduction of up to 30%. Several companies undertake process automation and data analytical innovations to boost operational productivity and waste minimization practices.

8.4.3   Reuse

Reusing objects means using them many times or reassociating them with new purposes before sending them to disposal. Products have longer lifespans through this strategy which produces lower waste totals.

8.4.3.1   Recent trends

A momentum shift toward the reuse movement has emerged from customer demand for sustainable products. Sharing economy platforms offering clothing as well as tool and electronic goods have shown substantial growth because people want to minimize their environmental effects. New studies show that by 2025 the secondhand market will expand beyond $64 billion worldwide. Local initiatives like swap events and repair cafés build community connections that motivate people to repair their possessions instead of throwing them out.

8.4.4   Recycling

The practice of recycling transforms waste materials into fresh products through an approach which saves materials while decreasing landfill waste levels.

8.4.4.1   Current challenges and innovations

The recycling field faces multiple obstacles which stem from materials of divergent quality and the continuous presence of contamination. The development of technology remains the primary objective in increasing recycling efficiency during this present era. AI-driven sorting technology installed at recycling facilities now helps operators achieve higher accuracy rates

while expanding recyclables recovery metrics. Multiple urban areas in 2023 implemented projects to address the recycling challenges of problematic waste items particularly flexible packaging and select plastics along with examining contamination factors and recycling stream potential growth (Kassab *et al.*, 2023).

8.4.5   Disposal

Disposal functions as a waste management option of last resort for materials which fail to meet requirements through other waste management processes. Two primary waste disposal methods are landfilling and incineration.

8.4.5.1   Innovations and improvements

Current disposal techniques put emphasis on both environmental stewardship and ecological safety. The technology at waste-to-energy facilities improves every year to turn unusable waste into usable energy more effectively. Some facilities will start integrating next-generation emission control systems featuring carbon capture technology to reduce greenhouse gas emissions following 2024 implementation. The latest technology used for landfills combines superior waste containment strategies and sophisticated leachate management techniques which reduces their impact on the environment.

8.5   Waste Reduction and Resource Recovery

Modern industrial waste management concentrates on waste reduction and material recovery, and thus it can become sustainable while minimizing waste production. In 2022, the EPA reported a 30% annual reduction in waste from all sectors of industry, thanks to efficient biodesign principles that enhance materials used to extend products' use.

Training and participation of employees are the backbone of the achievement of waste management goals. The IoT forms a basis for the actual time tracking of waste production, while artificial intelligence has predicted a 20% reduction in production waste by 2023.

The primary recovery alternatives that include recycling, composting, and waste to energy provide an important role to waste management. In the USA, 35% of the waste stream is recycled, a diversion of about 92 million tons of material away from the landfill each year; among the 25 million tons of that figure, organic matter has been removed through composting efforts. In 2023 alone, there were to be 77 waste-to-energy plants expected to turn 30 million tons of waste into more than 14 billion kilowatt-hours of electricity.

These strategies contribute to mitigating climate change and may produce significant cost savings of about 20%. Hence public engagement is critical to ensuring effective waste management in the shift to a circular economy.

8.6 Effective Waste Management Strategies

Today, modern waste management systems incorporate technological innovations centered on resource recovery while improving the position of environmental concerns. The circular economy's principles encourage repair, reuse, and product-as-a-service models for controlling e-waste. In 2023, new guidelines were set forth by the EU, promoting eco-design of durable, environmentally efficient products.

Emerging technologies such as chemical recycling, which deconstructs plastics into molecules for the creation of new products, appear to play a role in the recovery of materials. Blockchain technology boosts recycling transparency; as seen in pilot programs initiated in 2023, validating recycling practices.

Connected through the Internet of Things, smart trash bins optimize urban waste collection, reducing service travel distance by up to 15%. Data-driven waste audits have helped companies reduce waste generation by 25%. Stronger extended producer responsibility (EPR) legislation strengthens environmental outcomes by holding producers accountable for their end-of-life products. While community education initiatives can enhance recycling participation among cities by 40%, they simultaneously back city policies for zero waste overall with active community engagement.

8.6.1 Treatment and disposal methods

Different treatment and disposal methods enable effective waste management as they fight environmental threats and establish sustainable practices. Mechanical biological treatment (MBT) systems merge sorting technology with biological treatment approaches to eliminate between 30 and 70% of landfill waste according to the results of recent research conducted in 2023. US municipal solid waste thermal processing through incineration handles around 12% of total waste while producing more than 14 billion kWh of renewable energy in 2023. When subjecting organic materials to pyrolysis under oxygen-free conditions, operators gain access to bio-oil and gas while accomplishing plastic waste reduction. Advanced oxidation processes (AOP) employ powerful oxidants to handle dangerous waste materials and show confirmed effectiveness for water table cleanup. When it comes to waste disposal, authorities in the United States disposed of half of the 292 million tons of municipal solid waste generated during 2021 through landfill practices despite contemporary landfills possessing environmental protection systems. Secured landfills combined with deep well injections operate as disposal techniques for hazardous waste but face growing regulatory oversight. Standard organic waste management programs succeeded in removing about 25 million tons of organic materials from landfills in 2022 and the rising agricultural biosolid land applications in 2023 demonstrate a growing industry-wide acceptance strategy. Effective waste management and sustainability advances are necessary to achieve their mutual objectives.

8.6.2 Integrated waste management systems

Integrated waste management (IWM) means the waste minimization and sustainable urban management of environmental, economic, and social factors that include the handling, treatment, and disposal of waste. An IWM model will allow cooperation to take place from government bodies, private sectors, and communities with the aim of providing sustainable and effective waste management in combating a multitude of waste types and forms.

Developments surrounding IWM have in recent times gained momentum with the realization of the limited success that traditional methods of waste management, with their dependence on landfilling, present. More waste reduction and resource recovery would result when the complementary strategies of recycling, compositing, waste to energy, and waste reduction are better integrated within waste stream management practices. A finding published at the start of 2023 indicated that cities who followed IWM strategies were able to divert more than 30% of misuse from landfills in comparison to general and conventional methods of disposal.

The recent technological interventions such as data analytics and IoT have also played a crucial role toward IWM. These technologies enable municipalities to recognize patterns of waste generation, optimize collection techniques, and enhance recycling rates. An example can be derived from a pilot project in Stockholm, Sweden, which worked to implement an IWM system where smart waste bins provided live data on the kind and volume of the waste collected. This led to a 25% drop in collection costs and ensured a 15% boost in recycling rates in the neighborhood concerned.

Internationally, organizations such as the UN Environment Programme (UNEP) support IWM as part of the Sustainable Development Goals. Looking at cases where approaches to sustainable waste management have succeeded in 2023, it pointed to countries like Japan and Germany, where IWM systems have manifested as substantially lower generation per capita of waste along with recycling rates, affirming the effectiveness of integrated approaches.

8.6.3 Compliance with regulations

In 2023, the European Union updated its waste management directives, aiming for 65% of municipal waste to be recycled by 2035. Some countries that efficiently integrated these regulations into their IWM systems have made great strides; for instance, Germany, thanks to its stringent recycling laws, recorded a recycling figure of 67% in 2022, underlining the importance of regulatory frameworks. Not only does

compliance with such directives assure that the laws are followed, but it also fosters public trust and participation in waste management programs. Technological innovations have also contributed to improved compliance with the regulations. Cloud-based waste tracking systems offer opportunities for municipalities to monitor waste streams and report recycling rates in real time so that they may be in compliance with regulatory requirements. A prominent example is the implementation of an automated reporting systems in California that has aided compliance with the state's organic waste recycling mandate. By 2023, cities that employed such technologies reported 85% increases in rates of compliance. Training and community education are very much part of the compliance strategies. Recent ordinances have pointed toward a community-based approach to encourage empowered citizen action for waste reduction and recycling. As reported, cities that offered educational outreach had a 40% increase in community participation in 2023, directly correlating with improvement in compliance rates with local waste regulations. Overall, adherence to the regulations is the axle on which the wheel of IWM will turn. Through the incorporation of regulatory frameworks, technology, and community engagement, municipalities may institute sustainable waste management practice that is environmentally responsible and promotes public health.

8.6.4 Safety measures

Safety measures are of paramount importance in the waste management sector for the protection of the workers, communities, and the environment. Certain waste is rendered hazardous because of its inherent nature. Such hazardous wastes are handled, transported, and disposed of by following specific guidelines to reduce risk. In a 2023 survey undertaken by the Environmental Protection Agency (EPA), around 70% of waste management facilities enhanced their emphasis on safety standards in relation to increasing incidences of labor accidents and health issues. Among the major safety measures put in place are the provision of personal protective equipment (PPE), rigorous training measures, and operational policies aimed at preventing accidents. Audit of safety procedures is also, nowadays, very important to identify potential hazards. A 2022 study advised that the facilities conducting safety audits regularly had reduced workplace injuries by 40%, thus conclusively demonstrating that proactive safety strategies work.

Innovation also plays a crucial role in enhancing safety in waste management. For instance, the use of automated robotic waste sorting systems decreases the need for manual sorting, thus protecting workers from hazardous materials. Furthermore, the campaigns on public awareness are extremely important to safety in waste management. The communities are called upon to understand the safe methods to dispose of their wastes, especially hazardous waste, in order to avoid environmental contamination. A 2023 initiative in several cities in the USA has educated residents on proper disposal practices, resulting in a 25% drop in illegal dumping incidences. Safety recognition and enhancement are vital for sustainable waste management, offering health and safety for workers and communities surrounding waste facilities.

8.6.5   Risk assessment and emergency preparedness

Risk assessment and preparedness are indispensable preventive measures in waste disposal management. The whole process identifies the threat, evaluates risks posed by these threats to specific operations, and prepares for emergency situations relating to handling and disposal. A report from the Global Environmental Management Initiative in 2023 emphasizes that such comprehensive risk assessments can reduce the risk of accidents and environmental incidents by 60%. Risk assessment involves systematic evaluations of waste characteristics, operational practices, and potential emergency situations. Facilities shall analyze chemical reactions, exposure limits, and failure of equipment, among others, to formulate adequate safety protocols. In Texas, a hazardous waste facility recently implemented a new risk assessment framework. This framework identified weaknesses in their methods of handling and brought about substantive changes in their operations aimed

at risk mitigation. This may involve response procedures and communication procedures for handling crises. Training drills and simulations are regularly held to ensure staff are equipped to cope with emergencies faced, for example, by spills, fires, or explosions. According to the US Occupational Safety and Health Administration (OSHA), facilities with comprehensive emergency plans achieve up to 50% faster emergency response. Community engagement in emergency preparedness is important. In New Jersey, community drills and workshops in 2023 improved coordination of local emergency response to hazardous waste incidents, increasing preparedness while reducing public anxiety. Through comprehensive integrated plans, risk assessments, emergency preparedness, safety plans, etc., waste management facilities can enhance safety outcomes dramatically while protecting personnel and the environment.

8.7   Case Studies

The case study examples presented in this subject provide a group of innovative methods for the promotion of recycling and reduction of landfill space. What these case studies convey is that communities worldwide can adopt local solutions, motivational and technological, depending on the satisfaction of their residents.

8.7.1   Successful practices in waste management

The analysis shows that sustainable practices in waste management constitute a blueprint for the implementation of readily adoptable waste management strategies around the world. That is, for example, the so-called 'zero-waste' policy, which was formally adopted in Italy in Capannori, which was then a small town, in 2007. By introducing a pay-as-you-throw model that charges residents for the amount of waste they produce, Capannori managed to cut down on the waste sent to landfills by more than half in just some five or so years. Nowadays, Capannori has made claims of approximately 64% recycling, proving that economic incentives positively contribute to the promotion of sustainable behavior among residents.

Another successful example is Seattle, Washington, a highly active composting city since 2009. It made organics recycling mandatory, thus reducing solid waste by 20%. Seattle diverted over 200,000 tons of organic waste every year as of 2023. The composting promotion campaign in the city has also built community participation through workshops and incentives for backyard composting, with public participation growing substantially.

The city of Incheon in South Korea has introduced an advanced waste tracking system based on radio frequency identification (RFID) technologies to administer waste in an efficient manner. It is a real-time data provision on the disposal manner at waste disposal sites that serves as guidelines for implementing relevant waste management methods. This data facilitated the detection of additional recycling opportunities, which helped boost the recycling efficiency in Incheon from 2021 to 2023 to the tune of 30%.

The diversity of these case studies gives an example of how these creative practices could, in some parts of the world, serve as a vehicle for waste management and lead to some good results in the reduction of waste with the longevity of a community in mind.

8.8   Innovative Technologies

Innovative technologies in waste management, such as AI-driven sorting and IoT-enabled monitoring systems, enhance operational efficiency and safety. These advancements facilitate better recycling practices and optimize waste collection, contributing to sustainable management and reduced environmental impact in communities.

8.8.1   Advances in waste management and safety

Modern technological applications in waste management operations create improvements that increase safety standards as well as operational effectiveness. The integration of automation and advanced data collection systems and recycled processing methods are major priorities because artificial intelligence and machine learning systems are reshaping waste sorting operations. The waste management specialist AMP Robotics uses robots to accomplish recyclable sorting operations at a pace surpassing human abilities. In 2023 AMP Robotics formed an agreement to improve sorting precision by 30% which led to lower material contamination rates.

Real-time data collection from smart bins through IoT sensors allows collection services to improve their operation efficiency and minimize costs. The solutions implemented in Barcelona through IoT advancements during 2023 resulted in both decreased collection expenses by 25% and decreased environmental emissions.

Drone monitoring systems offer real-time surveillance that aids in workplace safety by warning staff about dangerous situations in landfill sites. Technological facilities implementing them show lower accident statistics thus proving their essential role in creating safer sustainable waste management procedures.

8.9   Promoting Sustainability

8.9.1   Fostering an eco-friendly corporate culture

An organization that wishes to exude an eco-friendly corporate culture in tandem with promoting sustainability in every respect must embark on quite a roundabout approach. The process starts where leaders pledge their commitment as executives who champion sustainable practices and set measures for success. For example, Unilever has structured sustainability into its core business, promising to lessen environmental impact while increasing social purposes to within a financially contoured result.

Employee engagement is critical for eco-friendly culture. Organizations increasingly create green teams or sustainability champions: volunteers who run programs such as recycling, energy conservation, and community outreach. In a 2023 study, those companies with green teams that were engaged witnessed a 15% increase in employee morale and productivity.

An education and training agenda should also remain ongoing. Companies can make sustainable practices a focus of the ongoing learning task by holding training workshops on waste reduction and energy efficiency, therefore empowering employees to adopt environmentally friendly decisions both in their workplace and at home.

Incentive programs further promote sustainability by incentivizing employees to live greener lives. Companies are offering bonuses or recognition as a reward for reaching sustainability goals. By embedding these principles in their culture, organizations can develop a mobilized workforce to drive long-term environmental change.

8.10   Conclusion: Summary and Recommendations

That sustainable waste management is a necessity is clearer than ever. As communities face waste generation increases and environmental challenges, embedding innovative technologies, using strategic case studies, and adhering to corporate sustainability become crucial ingredients for success.

In conclusion, corporations must foster their cultures of sustainability through strong employee involvement in developing sustainable practices. Investments should be made in training and support initiatives for sustainability of these efforts. AI and IoT solutions are enabling all stakeholders to execute these practices more efficiently with a clear focus on safety.

Sustainable management and cooperation among stakeholders, including government agencies, businesses, and the community will ensure that such strategies are adopted for decades into the future. Continuous evaluation and transparency in the reporting of sustainability goals should be implemented to ensure accountability and continued engagement.

This, in conclusion, includes initiatives serviced toward public awareness regarding waste management and instilling a culture of responsibility in people concerning waste management. Sustainability and correct practices will lead toward a cleaner, greener tomorrow. Thus, waste management will not only improve results, but communities and institutions will come together in protecting their environment.

References

Ahmed, S.F., Kabir, M., Mehjabin, A., Oishi, F.T.Z., Ahmed, S. *et al.* (2023) Waste biorefinery to produce renewable energy: Bioconversion process and circular bioeconomy. *Energy Reports* 10, 3073–3091. DOI: 10.1016/j.egyr.2023.09.137.

Alsabt, R., Alkhaldi, W., Adenle, Y.A. and Alshuwaikhat, H.M. (2024) Optimizing waste management strategies through artificial intelligence and machine learning – An economic and environmental impact study. *Cleaner Waste Systems* 8, 100158. DOI: 10.1016/j.clwas.2024.100158.

Ankit, S., Kumar, L., Tiwari, V., Sweta, J., Rawat, S. *et al.* (2021) Electronic waste and their leachates impact on human health and environment: Global ecological threat and management. *Environmental Technology & Innovation* 24, 102049. DOI: 10.1016/j.eti.2021.102049.

Bhattacharjya, S., Das, S. and Amat, D. (2021) Potential of microbial inoculants for organic waste decomposition and decontamination. In: Rakshit, A., Meena, V.S., Parihar, M., Singh, H.B. and Singh, A.K. (eds) *Biofertilizers*. Elsevier, Amsterdam, pp. 103–132. DOI: 10.1016/B978-0-12-821667-5.00027-0.

Chen, W., Xie, Y. and He, K. (2024) Environmental, social, and governance performance and corporate innovation novelty. *International Journal of Innovation Studies* 8(2), 109–131. DOI: 10.1016/j.ijis.2024.01.003.

Czekała, W., Drozdowski, J. and Łabiak, P. (2023) Modern technologies for waste management: A review. *Applied Sciences* 13(15), 8847. DOI: 10.3390/app13158847.

Debrah, J.K., Vidal, D.G. and Dinis, M.A.P. (2021) Raising awareness on solid waste management through formal education for sustainability: A developing countries evidence review. *Recycling* 6(1), 6. DOI: 10.3390/recycling6010006.

Fatimah, M., Qyyum, M.A., Lee, M., Alshareef, R.S., Aslam, M. *et al.* (2024) Industrial waste gases as a resource for sustainable hydrogen production: Resource availability, production potential, challenges, and prospects. *Carbon Capture Science & Technology* 12, 100228. DOI: 10.1016/j.ccst.2024.100228.

Fennelly, L.J. and Perry, M.A. (2018) *150 Things You Should Know About Security*. Elsevier, Oxford. DOI: 10.1016/B978-0-12-809485-3.00001-6.

Gebrekidan, T.K., Weldemariam, N.G., Hidru, H.D., Gebremedhin, G.G. and Weldemariam, A.K. (2024) Impact of improper municipal solid waste management on fostering One Health approach in Ethiopia – challenges and opportunities: A systematic review. *Science in One Health* 3, 100081. DOI: 10.1016/j.soh.2024.100081.

Groh, K.J., Backhaus, T., Carney-Almroth, B., Geueke, B., Inostroza, P.A. *et al.* (2019) Overview of known plastic packaging-associated chemicals and their hazards. *Science of the Total Environment* 651, 3253–3268. DOI: 10.1016/j.scitotenv.2018.10.015.

Kareem, H.A., Riaz, S., Sadia, H. and Mehmood, R. (2023) Industrial waste, types, sources, pollution potential, and country-wise comparisons. In: Riaz, U., Iqbal, S. and Jamil, M. (eds) *Waste Problems and Management in Developing Countries*. Apple Academic Press, Palm Bay, FL, pp. 169–203. DOI: 10.1201/9781003283621-8.

Kassab, A., Al Nabhani, D., Mohanty, P., Pannier, C. and Ayoub, G.Y. (2023) Advancing plastic recycling: Challenges and opportunities in the integration of 3D printing and distributed recycling for a circular economy. *Polymers* 15(19), 3881. DOI: 10.3390/polym15193881.

Kumar, A., Thakur, A.K., Gaurav, G.K., Klemeš, J.J., Sandhwar, V.K. *et al.* (2023) A critical review on sustainable hazardous waste management strategies: A step towards a circular economy. *Environmental Science and Pollution Research* 30(48), 105030–105055. DOI: 10.1007/s11356-023-29511-8.

Mattson, K.R., Pettersen, J.B. and Brattebø, H. (2024) Incineration economy: Waste management policy failing the circular economy transition in Norway. *Resources, Conservation and Recycling* 210, 107838. DOI: 10.1016/j.resconrec.2024.107838.

Mehta, P., Chelike, D.K. and Rathore, R.K. (2024) Adsorption-based approaches for exploring nanoparticle effectiveness in wastewater treatment. *Chemistry Select* 9(25), e202400959. DOI: 10.1002/slct.202400959.

Padmanabhan, K.K. and Barik, D. (2019) Health hazards of medical waste and its disposal. In: Barik, D. (ed.) *Energy from Toxic Organic Waste for Heat and Power Generation*. Elsevier, Duxford, pp. 99–118. DOI: 10.1016/B978-0-08-102528-4.00008-0.

Ramos, A. (2024) Sustainability assessment in waste management: An exploratory study of the social perspective in waste-to-energy cases. *Journal of Cleaner Production* 475, 143693. DOI: 10.1016/j.jclepro.2024.143693.

Sánchez-García, E., Martínez-Falcó, J., Marco-Lajara, B. and Manresa-Marhuenda, E. (2024) Revolutionizing the circular economy through new technologies: A new era of sustainable progress. *Environmental Technology and Innovation* 33, 103509. DOI: 10.1016/j.eti.2023.103509.

Shahbaz, M., Rashid, N., Saleem, J., Mackey, H., McKay, G. *et al.* (2023) A review of waste management approaches to maximise sustainable value of waste from the oil and gas industry and potential for the State of Qatar. *Fuel* 332, 126220. DOI: 10.1016/j.fuel.2022.126220.

Singh, B.J., Chakraborty, A. and Sehgal, R. (2023) A systematic review of industrial wastewater management: Evaluating challenges and enablers. *Journal of Environmental Management* 348, 119230. DOI: 10.1016/j.jenvman.2023.119230.

Singh, M., Singh, M. and Singh, S.K. (2024) Tackling municipal solid waste crisis in India: Insights into cutting-edge technologies and risk assessment. *Science of the Total Environment* 917, 170453. DOI: 10.1016/j.scitotenv.2024.170453.

Tihin, G.L., Mo, K.H., Onn, C.C., Ong, H.C., Taufiq-Yap, Y.H. *et al.* (2023) Overview of municipal solid wastes-derived refuse-derived fuels for cement co-processing. *Alexandria Engineering Journal* 84, 153–174. DOI: 10.1016/j.aej.2023.10.043.

Yu, H., Zahidi, I., Chow, M.F., Liang, D. and Madsen, D.Ø. (2024) Reimagining resources policy: Synergizing mining waste utilization for sustainable construction practices. *Journal of Cleaner Production* 464, 142795. DOI: 10.1016/j.jclepro.2024.142795.

9 Hospital Waste Management: A Case Study of Foud Al-Khateeb Hospital in Cox's Bazar

Md. Shafiqul Islam[1]* and Mohammad Abu Daud[2]

*[1]International University of Business Agriculture and Technology, Bangladesh;
[2]State University of Bangladesh, Bangladesh*

Abstract

Hospital waste is classified as both infectious and hazardous, presenting significant risks to environmental health and necessitating specialized treatment and management before final disposal. The risks associated with hospital waste management can arise from both intentional and unintentional exposure. The objective of this study is to evaluate hospital waste management by examining various waste collection methods, disposal systems, treatment processes, and the environmental impact of hospital wastes. The research involved observational assessments of hospital waste collection, transportation, storage, and disposal systems. Both qualitative and quantitative data were gathered through direct field observations, focus group discussions, and data from secondary sources. Hospital generated highly toxic chemicals that pose a risk for disease transmission. Sharp and infectious wastes are hazardous for environmental and human health. There is a lack of a systematic approach to waste management as it does not utilize color-coded bins for waste disposal. Inadequate waste segregation facilitates pollutants proliferation and dissemination by insects, rodents, or environmental factors such as wind and precipitation. Liquid poses a risk of contaminating groundwater and entering the food chain. Implementing effective practices for the collection, segregation, storage, treatment, and disposal of hospital waste is essential for public health and the environment.

9.1 Introduction

Hospital waste is both infectious and dangerous, posing significant risks to environmental health. It necessitates specialized treatment and management before its ultimate disposal. Managing medical waste is an essential component of healthcare operations, which involves the appropriate handling, treatment, and disposal of waste produced in healthcare facilities (HCEs) (Lee and Lee, 2022). In developing nations such as Bangladesh, effective medical waste management becomes increasingly vital due to the significant prevalence of infectious diseases and the rising number of patients seeking medical attention (Rashbrook *et al.*, 2000). Poorly managed medical waste can pose serious risks to both public health and the environment (Ravindra *et al.*, 2023), making it crucial to evaluate current policies and practices to identify obstacles and create timely strategies for sustainable development. The global consensus acknowledges the importance of managing healthcare waste safely and appropriately.

*Corresponding author: dislam.env@iubat.edu

© CAB International 2025. *Municipal Waste Management: Policies and Strategies*
(eds S. Kumar *et al.*)
DOI: 10.1079/9781836990666.0009

Research has indicated that contact with dangerous medical waste can result in serious health problems, including infections, injuries, and chronic illnesses (Janik-Karpinska *et al.*, 2023). In addition, the pollution from untreated medical waste has the potential to contaminate water bodies, soil, and air, threatening ecosystems and contributing to environmental damage (Sonone *et al.*, 2020). The World Health Organization (WHO) and the Environmental Protection Agency (EPA) have highlighted the necessity of properly handling and disposing of hospital waste generated by healthcare facilities. In Bangladesh, city corporations are tasked with managing solid waste. Nevertheless, waste collection frequently occurs irregularly, resulting in unsanitary conditions near primary collection areas. Numerous towns lack the capacity to collect all waste, and proper sanitary disposal often presents challenges. The management of hospital waste is becoming increasingly important due to the rising number of hospitals, clinics, and diagnostic labs, particularly in Dhaka City. Consequently, healthcare waste management is often overlooked and falls under the jurisdiction of local municipal bodies responsible for the collection, removal, and disposal of various types of waste from public areas. Bangladesh have approximately 460 Upazilla-level hospitals and 9722 community-level clinics, alongside around 1449 outdoor health facilities at the union level, all under the Directorate General of Health Services (DGHS). According to the DGHS in 2010, there are about 117 hospitals currently in operation at the district level. Among private hospitals, there are about 2501 registered facilities and 5122 diagnostic centers nationwide. Additionally, there are numerous clinics, including around 5000 government and NGO-operated clinics and doctors' offices where healthcare waste is produced. In Dhaka City alone, there are roughly 1200 hospitals, clinics, and diagnostic centers (Mazumder *et al.*, 1997). According to a research report by the Dhaka City Corporation, the waste generated per person per day is approximately 0.5 kg. The same report identified that Dhaka City produces about 3700 metric tons of waste daily, of which around 200 tons is hospital waste and 40 tons is classified as infectious waste (Rahman and Melville, 2000). Limited understanding of how to separate hazardous from non-hazardous

waste makes it challenging to obtain accurate information regarding the actual daily waste output. The effective execution of medical waste management policies necessitates collaboration and proactive participation from various stakeholders, such as healthcare workers, waste management agencies, legislators, and the surrounding community (Pereno and Eriksson, 2020). This study aims to examine the hospital waste management practices at Fouad Al-Khateeb Hospital, along with the reasons behind management failures and their environmental impacts in that area. Improper disposal of waste generated in healthcare settings can directly affect the health of the community, the staff working in these facilities, and the surrounding environment. Waste from kitchens, packaging materials, and waste paper are generally considered general wastes produced in hospitals and clinics due to administrative or housekeeping functions. Pathological wastes represents another category of hospital waste, involving tissues, organs, human fetuses, placenta, blood, and other bodily fluids. Such wastes are often classified as hazardous. Infectious wastes form another category of hospital waste, consisting of pathogens. These wastes must be handled with particular caution to prevent the spread of infections from their management. Tissue cultures, stocks of infectious agents in laboratories, surgical waste, and waste from infectious patients are classified as infectious wastes. Sharps, which include needles, broken glass, saws, nails, blades, and scalpels, are also part of hospital waste. Pharmaceutical waste includes products, drugs, and chemicals that are outdated or contaminated, forming a specific type of hospital waste. Chemical wastes consist of discarded solids, liquids, gaseous chemicals, disinfectants, and reagents, which are hazardous, as are radioactive wastes such as those derived from radio-nuclides for tumour localization or therapeutic procedures for cancer-related diseases, and so forth. Healthcare organizations are accountable for keeping all personnel, from nurses to waste handlers, informed to ensure safe management of hazardous waste: absence of oversight, absence of regulations, application of the '3Rs policy', and the 'polluter pays principle' (Shammi *et al.*, 2022).

9.1.1   Objectives

The primary aim of this study is to evaluate the current status of hospital waste (solid, liquid, hazardous, and non-hazardous) management at Fouad Al-Khateeb Hospital concerning appropriateness, problems, and challenges encountered.

The specific objectives of this study are to:

1. Evaluate different types of hospital solid waste standard collection procedures and their standard disposal systems.
2. Assess improved alternative treatment procedures for hospital liquid waste in a compact setting.
3. Make an assessment of environmental and health impact of hospital waste.
4. Identify the problems and challenges of hospital waste management.
5. Recommend the strategies for potential solutions.</nl>

9.2   Literature Review

The World Health Organization (WHO, 2018) described medical waste as trash produced during medical procedures and activities at different healthcare facilities, such as clinics, hospitals, diagnostic centers, research facilities, and laboratories. The use of medical waste management (MWM) services could be an important way to produce bioenergy potential, linked to factors like poverty alleviation, equity, inclusivity, and food security, as well as their overall contribution to achieving the Sustainable Development Goals (Edwards et al., 2021). Medical waste management has been turned into a driving force behind traditional economic growth in wealthy nations (Kandasamy et al., 2022). However, in low-income nations like Bangladesh, Sri Lanka, Myanmar, India, and Pakistan, there is still a dearth of empirical research looking at comprehensive ways to improve MWM services (Khan et al., 2019). Furthermore, it has been difficult for these studies to provide cohesive and useful therapies. This difficulty arises because, in many unstable low-income nations, significant resources are devoted to providing technical and financial assistance for MWM systems; nevertheless, the success of these initiatives depends on how they are implemented in the specific environment (Kane et al., 2019).

9.2.1   Sources, types, and management of hospital waste

Hospital garbage, also referred to as medical waste, is the term used to describe the waste produced by hospitals, clinics, or other healthcare facilities. Hospitals generate a wide range of medical waste, from dangerous biological, chemical, and radioactive wastes to ordinary garbage (such as food and paper waste) (WHO, 2018). Medical wastes can be broadly divided into two categories (Table 9.1). The pattern, types, and composition of medical wastes differs from country to country; even within a country it depends on the type of health services and which providers are providing what to the people. Pruss et al. (1999) reported on different types of hospital waste in the world. There are two distinct sources of medical waste, namely, major and minor. The minor sources are scattered and produce similar kinds of waste as major sources, but these waste does not contain radioactive wastes, no human body parts, or sharps items like needles (WHO, 2018).

9.2.2   Effects of hospital wastes

Medical wastes contain infectious agents, genotoxic materials, toxic or hazardous chemicals or medications, and sharps germs that can enter the human body through a variety of routes and cause diseases. Improper handling of these wastes clearly poses a risk to both humans and the environment. Injuries such as radiation burns or injuries from sharp objects can also result from waste and by-products. Poisoning and pollution can occur from the release of pharmaceutical products, especially antibiotics and cytotoxic drugs, through wastewater and toxic elements (WHO, 2000). Infectious disorders of the stomach, lungs, urinary system, skin, meningitis, viral hepatitis, and tetanus are the major illnesses. Numerous other illnesses could be spread by coming into contact with medical waste. Many people are at risk both directly and indirectly because they handle medical waste

Table 9.1. Sources of different types of hospital waste. Author's own

Major sources		Minor sources		Reference
#	Name and source	#	Name and source	
1	Hospitals University hospitals General hospitals District hospitals Upazila hospitals Private hospitals Clinics	1	*Small healthcare establishment* Physicians' offices Dental clinics Acupuncturists Chiropractor	Prus *et al.*, 1999WHO, 2018
2	Other healthcare Emergency medical services Healthcare centers and pharmacy Obstetric and maternity clinics Outpatient clinics Dialysis centers Military medical services Long-term healthcare establishment	2	Specialized healthcare establishments and institutions with low waste generation Convalescent nursing homes Psychiatric hospitals Disabled persons' institutions	Prus *et al.*, 1999 WHO, 2018
3	Laboratories and research centers	3	*Non-health activities involving intravenous* Cosmetic ear piercing and tattoo parlor Illicit drug users	Prus *et al.*, 1999 WHO, 2018
4	Mortuary and autopsy centers	4	Funeral services	Prus *et al.*, 1999 WHO, 2018
5	Animal research and testing	5	Ambulance services	Prus *et al.*, 1999 WHO, 2018
6	Blood banks blood collection services	6	Home treatment services	Prus *et al.*, 1999 WHO, 2018

and come into contact with it. Hospital wastes, especially hazardous waste, creates health risks because it contains infectious agents, toxic and hazardous materials, and sharp wastes. Healthcare personnel and waste workers (collectors, segregators, and transporters) come into contact with such types of hazardous waste and can be infected with various diseases. Many hazardous wastes are disposed without treatment and many people are affected from landfills. Hospital wastes cause occupational risk, public health risk, and risks to the environment.

9.2.3   Medical waste management

In general, waste management refers to the process of controlling garbage from the point

of generation to the final disposal. Hospital waste management is the efficient separation of waste and the distinct treatment and disposal of each waste category, which would not have been possible without the dedication of senior directors and the inspiration of the medical and support workers (Rashbrook *et al.*, 2000). Nowadays, managing medical waste is a top priority for both developed and developing nations worldwide. The main goal today is to manage healthcare waste in a way that is both economically and environmentally feasible for all nations, but particularly for developing nations. Enhancing healthcare waste management at the local, regional, and national levels requires the creation of goals and the planning necessary to attain them (WHO, 1998). The Agenda 21 suggests for appropriate waste management that include prevention and reduce the development of trash; to reuse or recycle as much as possible; and to dispose of waste via final residues in landfills that are carefully planned and constrained (Rashbrook *et al.*, 2000). Therefore, avoiding health risks and creating a hospital environment that is conducive to well-being are prerequisites for effective hospital waste management. Guidelines for safe medical waste management were established by the World Health OrganizationAkther (2000) and should be followed by all institutions. These include: (i) minimizing waste at the source through recovery, reuse, and stock management; (ii) separating waste into different categories and sharps; (iii) identifying waste by colour coding it; (iv) collecting and storing waste through regular programs; (v) transferring waste by using enclosed vehicles and the 'chalked path' method from generation to disposal sites; and (vi) treating waste using both burn and non-burn methods. Effective medical waste management is a relatively new phenomenon in Bangladesh, and the government is working to create a fresh, cutting-edge strategy to handle medical trash. Bangladesh has no national policy on the management of medical waste. The government should prioritize developing a framework for the appropriate and scientific handling of medical waste. The *Hospital Waste Management Pocket Book* was created by the Department of Environment in 2004 and updated in June 2010 (DGHS (Directorate General of Health Services), 2008). Medical Waste ManagementDGHS (Directorate General of Health Services), 2008 (DGHS, 2010) support this book. All standard procedures for an appropriate management system for all healthcare facilities are included in the pocket book. Nearly all hospitals and clinics dispose of both hazardous and non-hazardous garbage in the local dustbins or by the side of the road without any kind of treatment. It was found that some hospitals separate infectious and noninfectious waste streams at the point of production, the wastes are mixed together in municipal dustbins for disposal (Hassan *et al.*, 2008). The typical hospital waste management system in Bangladesh is discussed in a report on waste concerns (2002).

9.2.4 Reasons for failure of hospital waste management

Hospital waste management in Bangladesh needs to be stressed as it is causing serious damage to health and environment. Despite having a hospital waste management pocket book since 2008, the government has not been successful to ensure the implementation of the guideline in both public and private hospitals/clinics. Hospitals have no interest and are not bothered to improve or update their disposal methods as there is a cost involved. Institutions that are aware do not have proper management systems and guidelines. However, the actual problems observed in both public and private hospitals/clinics and diagnostic centers are (i) no alternative methods for safe disposal; (ii) no system for segregating the waste before disposal; (iii) no specific regular awareness program among all staff; (iv) no protection for waste handlers, which are often infectious and potentially dangerous; (v) no specific training program for the nurses and cleaners regarding waste handling, disposal, or management. Problems include: (i) lack of implementation of guideline; (ii) existing gaps within the DGHS (Directorate General of Health Services) (2008); (iii) problem in the Environmental Act 1995; (iv) lack of interest and unity; (v) economic constrains; (vi) following previous management; (vii) corruption of the lower level; (vii) hazardous waste management is not high in the political agenda; (ix) lack of responsibilities and supervision; and (x) inadequate enforcement of existing pollution control laws. According to Mubarak (1998), private hospitals and clinics are better than public hospitals/clinics because in public hospitals autocracy is practiced where policy makers are not interested to improve the hospital environment and follow the waste management procedures even if they are aware of it. All are lacking from awareness on occupational health and safety (Dana, 1999). These interventions will aim to tackle particular obstacles related to segregation, transportation, storage, and disposal services (Rahman and Saidur, 1999). Hospital waste management policy and laws was mentioned by a study of WHO (2018). The author compiled different policies and laws in chronological order (Tables 9.2).

Table 9.2. Chronology of policies and laws of hospital waste management in Bangladesh. Author's own

Name of the act	Year	Reference
Private Hospitals and Laboratories	1982	WHO, 2018
National Environmental Policy	1992	WHO, 2018
The Environment Conservation Act 1995	1995	WHO, 2018
Environment Conservation Rules	1997	WHO, 2018
Environment Court Act	2000	WHO, 2018
Medical Waste Management and Processing Rules	2008	WHO, 2018
National Guideline for Medical Waste Management	2016	WHO, 2018
Private Hospital, Clinics and Diagnostic Centers Administration Rules	2016	WHO, 2018

9.3 Methodology

This study focused on the present status of hospital waste management practice at Foud Al-Khateeb Hospital. Data collection included hospital waste collection, transportation, and storage and disposal systems. This study identified the lack of waste management and the authority's future management plan. Both qualitative and quantitative data were collected through direct field observations, focus group discussions with the stakeholders and employees of the organization; secondary information was also collected for proper documentation, such as research articles, books, and periodicals. Both primary and secondary sources was used to collect data as fulfilment of the study. The study followed the following steps:

1. First, the existing practices of hospital waste management were identified. Interviews were held with healthcare workers and the departmental head about the present practices of hospital waste management system.
2. Second, the program for minimization, prevention of clinical waste were designed to observe the proper segregation using advanced technologies.
3. The third stage included examining the possibilities for communication of the existing system of hospital waste with the present municipal system for the collection, transportation, processing, and disposal of solid waste. It is important to note that clinical waste should not be considered separately from the city complex system of waste management. The study examined how to develop hospital waste collection, transportation, processing, and disposal in relation to standard procedure.
4. The fourth stage was to examine the monitoring and enforcement system for proper management of hospital solid and liquid management by the Department of Environment (DoE) of the district.
5. Finally, the ways to build awareness and control in decision making and management of hospital waste were assessed and recommendations were made. At this stage the existing system of control over the implementation of legislation was investigated and proposals for its improvement were developed with experience of Dhaka district.

Thus, the review of the legislation of the hospital and documents were carried out. Interviews with key personnel involved in the management of hospital waste provided the initial information for analysis. Observation was made to know the present practices, including procedures for the appropriate handling, routing, and disposal. Two focus group discussions (FGDs) were carried out with the organization and with an MWM-related person and workers to get proper information. Five key informants interviews were conducted with DoE authority of Cox's Bazar, a representative of the Civil Surgeon's office, a representative of the BCC, the president of Cox's Bazar Civil and Structural Engineering Association, and an admin officer of the World Food Programme (WFP) under the UN project.

Table 9.3. Waste category and description with examples.

Waste category	Description with examples
Infectious waste	Pathogens present in this type of waste including. excreta, laboratory cultures, tissues, materials, or equipment that have been in contact with infected patient.
Pathological waste	Human tissues or fluids including blood and other body fluids, fetuses, organs, placentas, and tissue.
Pharmaceutical waste	Wastes containing pharmaceuticals such as pharmaceuticals that are expired.
Genotoxic waste	Waste containing substances with genotoxic properties including cytotoxic drugs that used in cancer therapy and genotoxic chemicals.
Chemical waste	Chemical substances present in a waste. Such as reagents, film developer; disinfectants solvent dressings, stool napkins, and plaster cast.
Wastes with high content of heavy metals	Batteries, broken thermometers, and blood-pressure gauges.
Pressurized containers	Gas cylinders, aerosol cans. Radioactive waste such as unused liquids from radiotherapy or laboratory research, contaminate glassware, packages, or absorbent paper.
Sharps	Sharp wastes, including needles, knives, blades, broken glass infusion sets, syringes, and scalpels.

9.4   Results

9.4.1   Study area

The research focused on the Fouad Al-Khateeb Hospital, located on Central Mosque Road in Cox's Bazar, Bangladesh. This facility is among the largest private hospitals in Cox's Bazar. It serves as a multi-specialty healthcare service provider. The coordinates of the hospital are 21°26'30.53"N and 91°58'33.69"E. The hospital consists of seven stories and has a capacity of 100 beds as a multi-specialty institution.

9.4.2   Waste category and generation

Different categories of medical waste found at Fouad Al-Khateeb Hospital may contain highly toxic chemicals and can present a mechanism for transmission of diseases. The growth of the medical sector around the world over the last decade combined with an increase in the use of disposable cheap medical products of different category (Table 9.3) has contributed to the large amount of medical waste.

Sharp waste and infectious waste is hazardous for the environment as well as for humans and animals. Results showed that general wastes produced more in the hospitals followed by recycle waste, infectious waste, and sharp waste. Different types of medical waste generated from different department of the hospital. Data (Table 9.4) has been collected from the two FGDs with the hospital authority and related workers of the organization involved in medical waste handling procedures.

9.4.3   Standard system of waste collection

It was directed by the concerned authority to manage hospital waste following the standard of waste collection, segregation, transportation, storage, treatment, and disposal. According to guidelines, the collection of and segregation should be done using different color-coded bins. Black bins need to be used for general waste which is collected from kitchen and medicine boxes. Yellow bins are dedicated for infectious wastes which are collected from operation

Table 9.4. Waste generation from different departments. Author's own

SL	Name of department	Generated waste in kg. or liter/day				
		Sharp(in kg)	Infectious(in kg)	Liquid(in liter)	Recycle(in kg)	General(in kg)
01	OT	0.25	2.75	1000.00	0.25	1.25
02	Ward/cabin	1.50	3.50	0	5.50	10.50
03	OPD	0	0	500.00	8.50	15.00
04	Blood collection	0.05	1.50	5.00	3.00	1.00
05	Pathology	0.25	1.00	100.00	1.00	0
06	Kitchen	0	0	0	7.00	125.00
07	Washing plant	0	0	15,000.00	0	0
08	Hospital floor wash	0	0	140.00	0	0
Total		2.05	8.75	16,745.00	25.25	152.75

theaters, cabins, wards, intensive care units, pre- and post-operative rooms, such as cotton bandages, amputated body parts, placenta, and blood and urine bags. Blue bins are used to collect liquid waste from the laboratory, operation theaters as blood, laboratory machine discharge waste, and suction fluid. Sharp items use red bins which is collected from blood sample collection rooms, cabins, and wards, intensive care units, pre- and post-operative rooms, such as needles, blades, knifes, vial/ampoules. Green bins are used only for recyclable items which are collected from all areas of the hospitals, such as plastic bottles, syringe without needles, saline bag, and gloves. It was found that the personnel and workers who are responsible for this are not following the standard guidelines and used wrong bins for waste collection, segregation, and storage.

9.4.4 Disposal system of hospital waste

Guidelines instructed that black category (general waste) medical waste needed to be disposed to the City Corporation household collection dump. Yellow category (infectious waste) medical waste should be disposed to the City Corporation's infectious medical waste dump. City Corporation would incinerate the infectious medical waste at their specific area. All blue category (liquid waste) medical waste needed to be disposed to the hospital or central effluent treatment plant. And

after treatment of the liquid waste the treated water should be reused or disposed to the sewage line. Sharp items (red category) medical waste should be disposed to the hospital's own/central deep burial pit; and green category (recycle waste) medical waste must be recycled for reuse. Though the hospital had bins, they were not following the correct use for the different color-coded bins and did not segregate.

9.4.5 Problems and challenges of hospital waste management

Many problems and challenges were identified in this study related to medical waste generation, collection, segregation, and disposal procedures. Working staff and workers were lacking awareness and training on medical waste management systems. Improper recordkeeping of different types of waste generation was found. There was an absence of regular monitoring by the hospital authority. Proper monitoring by the government's authority was irregular. There was no facilities for short-term storage and a lack of sufficient City Corporation dumping bins, dumping stations, and incineration sites. There was no burial pit for the management of sharp medical waste. There was no coordination among authorities. The services were lacking from adequate manpower. Community awareness and participation was not present. Hospitals had no needle crusher or

Table 9.5. Satisfaction level on hospital waste management. Author's own

KII person	Satisfaction level for medical waste		
	Collection	Segregation	Disposal
DoE representative	Δ	↓	↓
Sanitary inspector	Δ	↓	↓
President, Cox's Bazar Civil and Structural Engineering Association	Δ	↓	↓
City Corporation representative	Δ	Δ	↓
Admin officer of WFP under UN project	↓	↓	↓

Poor Δ; Dissatisfying ↓

burner at hospital wards, cabins, operation theaters, or other premises.

9.4.6 Satisfaction level of the respondents on hospital waste management system

According to an interview with the representative of DoE, most hospitals were renting buildings for use and as a result they were not able to ensure compliance of the hospital waste management requirements. The building was not purpose-built for a hospital. Waste segregation and proper disposal were their main challenge. He mentioned that for liquid waste management a central wastewater treatment plant was needed for liquid waste management and to protect water pollution from medical hazardous waste. The president of Cox's Bazar Civil and Structural Engineering Association said that medical waste management is not satisfactory (Table 9.5). He said that City Corporation needed to designate separate dumping stations for medical waste final disposal. The sanitary inspector mentioned that they were trying hard to regulate the proper medical waste management system and failed to do this due to lack of sufficient manpower. He agreed to set up a central wastewater treatment plant for medical liquid waste management. The City Corporation's representative mentioned that they will establish separate medical waste dumping stations and pits for medical waste segregation. The admin officer of WFP under the UN project was annoyed because of the overall management procedure of medical waste. She requested authorities to develop the medical

waste management system and to protect the environment of the world's longest sea beach of Cox's Bazar. It was found that improper medical waste collection, segregation, and disposal caused environmental problems by increasing air and water pollution, decreased the ecosystem of the adjacent canal and rivers, and increased contamination due to improper disposal of the hazardous medical waste.

The organization had a wastewater treatment plant for their medical and sewage water treatment. It was appreciated that they did not directly throw their liquid waste into the sewage system. They were reusing the treated wastewater, however, if the biological process was considered, they needed an extra two–three tanks for the process, a primary clarifier tank, a secondary clarifier tank, and a flocculation tank, in addition, the biological treatment of 25,000 liters of wastewater required a minimum of $28\,m^2$ ($300\,ft^2$) space. It is common knowledge that the majority of private hospitals do not have enough space to build effluent treatment plants or wastewater treatment plants.

9.4.7 Effects of improper management of medical waste to the environment

It was found that the study hospital was trying to manage but it was neither appropriate nor enough. The negative effects that could result from the poor management of collection methods and the disposal of solid waste resulting from health institutions, whether in terms of the safety of the medical staff working within the institution or in terms of environmental safety in general, were numerous. The disposal of solid waste in terms of management and procedures might lead to the spread of pollutants in these wastes by insects, rodents, or wind, as well as rain or floods that might lead to seepage into soil or groundwater. The landfills of medical waste and the transfer of these pollutants to soil or seabed, caused a major threat to the environment and fisheries. Improper segregation of medical waste and different medical waste streams from the point of origin could trigger a domino effect on the environment which posed danger to people, animals, and soil and water sources. Improperly disposed of biomedical wastes caused lung infection, parasitic infection, skin infection, the spread of viral illnesses, cholera, and tuberculosis. Needle stick injury and sharps injury incidents were an increased risk to health (Table 9.3).

9.5 Discussion

In this study, it was found that medical waste management is not receiving priority attention. In contrast to first world countries where this issue has received significant attention and resources, many poor countries have not prioritized medical waste management, resulting in inadequate resources and improper methods being used (Ali *et al.*, 2017). Many developing countries have not made medical waste management a priority, which has led to inadequate resources and inappropriate procedures, in contrast to wealthier nations where this issue has received significant attention and resources (Ali *et al.*, 2017). Bangladesh is still lacking the implementation of medical waste policies. Implementation issues still exist even though neighboring nations such as India and Sri Lanka have changed their policies (Azam *et al.*, 2020). Lack of awareness and the level of medical waste management was very low. Despite these initiatives, it is important to remember that medical waste in Bangladesh is frequently disposed of similarly to regular household waste, underscoring the need for better management techniques (Rahman and Melville, 2000). Foud Al-Khateeb Hospital authority is not complying, even after receiving the DGHS guidelines for hospital waste management. According to Basak *et al.* (2019),

waste management in Bangladesh is not given enough consideration by the government or hospital administration. Collection and segregation is not taking place properly due to lack of knowledge and willingness in the hospital. Even though a study mentioned the carelessness, ignorance of the issue, and resistance to managing medical waste, the HCEs' management still did not prioritize segregation (Caniato *et al.*, 2015). More detailed information on medical waste disposal methods is required in order to ensure the best practices for medical waste disposal (Korkut, 2018). Researchers and industry experts could acquire important insights that would aid in the creation and application of better plans for the secure and effective handling of medical waste by acquiring more information (Baghapour *et al.*, 2018). According to an audit report from 2016, they failed to maintain the proper records of treated medical waste, including recyclable garbage. In a study on medical waste disposal in Brazil, Cesario *et al.* (2020) found that 66% of garbage was prevented from accumulating by using the segregation process, of which 95% was recyclable waste that might potentially have an effect on the nation's economy. A study on sustainable medical waste management in Bangladesh found that although the nation is making some progress in creating medical waste policies, there are still gaps in the country's current sustainable medical waste management practices because of ineffective and weak organizational structures, such as the creation of district and Upazila level committees (Barua and Hossain, 2021). In Bangladesh, one of the most overlooked aspects of waste management is the handling of medical waste.

Therefore, it is necessary for institutional leadership to take the initiative to implement policies (Sordi Schiavi *et al.*, 2021). Sordi Schiavi *et al.* (2021) highlight important aspects of building medical waste management leadership. Due to a lack of intersectoral coordination for sustainable programs, a shortage of qualified personnel, long-standing sociocultural norms, and related behavioral barriers, adopting such assistance in Bangladesh may take a while (Sujon *et al.*, 2022; Debnath *et al.*, 2023). Research revealed that the segregation, storage, and disposal of medical wastes were hindered by a lack of personnel (Chisholm *et al.*, 2021). Similarly, human resources represent the biggest obstacle in Bangladesh's health systems (Ahmat *et al.*, 2022).

9.6 Conclusion

Hospitals in Bangladesh pose a significant threat to health and environment on account of inadequate waste management and a need to raise awareness and provide educational training on medical waste management. It needs to bring innovative changes in healthcare waste management. However, the healthcare waste management guideline, planning, and policy should be under the shadow of legislation; emphasis should be placed on the development of educational training programs, recordkeeping, monitoring, and reviewing of the existing situation, and there should be collaboration between interministerial and hospital authorities, as well as active participation from the community. Indeed, a good number of organizations are working on a healthcare waste management system. Now the task is for government to formulate an appropriate policy collected in a manual that needs to be followed by all the hospitals, clinics, and diagnostic centers in Bangladesh. At the same time, the government and the private sector should create a central facility for safe collection and final disposal of the medical waste, through landfill and incineration.

9.7 Recommendations

Based on assessment the few recommendations were made for Fouad Al-Khateeb Hospital as well as all the hospitals of Bangladesh. The recommendations are applicable at the practice level, management level, and research level, and the policy issue of the country.

- Proper record of the different streams of medical waste generated should be kept.
- Formal training on MWM practices to nurses, ward workers, and the cleaners needs to be carried out.
- Used needles and syringes should be properly destroyed before disposal.
- Proper transportation of the medical waste from the wards by waste trolleys should be done.
- The use of personal protective equipment such as gloves, masks, and boots should be ensured.
- The licenses obtained by the healthcare facilities, i.e. an operating license from DGHS and a trade license from the City Corporation do not provide any cross linkages to the medical waste management, this would require policy interventions involving three different ministries.
- Timely collection, transportation, and disposal of medical waste should be done by the authority.
- The pollution control board should be very strict in nature and they should carry out inspections without any prior notice.
- The government should attempt to educate the people in public about the consequences of mishandling biomedical waste.
- The environmental science should be a compulsory subject to all the courses with the chapters of biomedical waste and its impact.
- Every hospital should have special color-coded bags to use in the bins for biomedical waste.
- Every hospital should have special closed-container vehicles to carry the waste.

Acknowledgment

The authors are thankful to the community, especially the respondents who provided relevant information during data collection.

References

Ahmat, A., Okoroafor, S.C., Kazanga, I., Asamani, J.A., Millogo, J.J.S. *et al.* (2022) The health workforce status in the WHO African region: Findings of a cross-sectional study. *BMJ Global Health* 7(Suppl 1), e008317.

Ali, M., Wang, W., Chaudhry, N. and Geng, Y. (2017) Hospital waste management in developing countries: A mini review. *Waste Management & Research* 35(6), 581–592.

Azam, R., Parveen, N., Singh, D.V. and Azam, Z. (2020) Generation and management of biomedical waste. In: Bhat, R.A., Qadri, H., Wani, K.A., Dar, G.H. and Mehmood, M.A. (eds) *Innovative Waste Management Technologies for Sustainable Development. IGI Global*. Hershey, PA, pp. 98–121.

Baghapour, M.A., Shooshtarian, M.R., Javaheri, M.R., Dehghanifard, S., Sefidkar, R. *et al.* (2018) A computer-based approach for data analyzing in hospital's health-care waste management sector by developing an index using consensus-based fuzzy multi-criteria group decision-making models. *International Journal of Medical Informatics* 118, 5–15.

Barua, U. and Hossain, D. (2021) A review of the medical waste management system at COVID-19 situation in Bangladesh. *Journal of Material Cycles and Waste Management* 23(6), 2087–2100.

Basak, S.R., Mita, A.F., Ekra, N.J. and Alam, M.J.B. (2019) A study on hospital waste management of Sylhet City in Bangladesh. *International Journal of Engineering Applied Sciences and Technology* 4, 36–40. DOI: 10.33564/ijeast.2019.v04i05.006.

Caniato, M., Tudor, T. and Vaccari, M. (2015) International governance structures for health-care waste management: A systematic review of scientific literature. *Journal of Environmental Management* 153, 93–107. DOI: 10.1016/j.jenvman.2015.01.039.

Cesario, F.K.O., Fontoura, R.P., Conceição, A.H., Jr, Cruz, A.G., Nimer, N.F.S. *et al.* (2020) Reduction of management costs and avoidance of air release of carcinogens through a waste segregation program in a Brazilian medical institution. *Frontiers in Public Health* 8, 583962.

Chisholm, J.M., Zamani, R., Negm, A.M., Said, N., Daiem, M.M.A. *et al.* (2021) Sustainable waste management of medical waste in African developing countries: A narrative review. *Waste Management & Research* 39(9), 1149–1163.

Concern, W. (2002) Manual on bio-medical waste management. Unpublished ms.

Dana, T. (1999) Hospital waste disposal an exploration in search of policy, guidelines and rules. In: *Bangladesh Legal Aid and Services Trust (BLAST)*. Dhaka, Bangladesh.

Debnath, B., Bari, A.M., Ali, S.M., Ahmed, T., Ali, I. *et al.* (2023) Modelling the barriers to sustainable waste management in the plastic-manufacturing industry: An emerging economy perspective. *Sustainability Analytics and Modeling* 3, 100017.

DGHS (Directorate General of Health Services) (2008) *Medical waste management rules 2008*. DGHS, Bangladesh.

DGHS (2010) *Hospital Waste Management Pocket Book*. DGHS, Bangladesh.

Edwards, D.P., Cerullo, G.R., Chomba, S., Worthington, T.A., Balmford, A.P. *et al.* (2021) Upscaling tropical restoration to deliver environmental benefits and socially equitable outcomes. *Current Biology* 31(19), R1326–R1341.

Hassan, M.M., Ahmed, S.A., Rahman, K.A. and Biswas, T.K. (2008) Pattern of medical waste management: Existing scenario in Dhaka City, Bangladesh. *BMC Public Health* 8, 36. DOI: 10.1186/1471-2458-8-36.

Janik-Karpinska, E., Brancaleoni, R., Niemcewicz, M., Wojtas, W., Podogrocki, M. *et al.* (2023) Healthcare waste – A serious problem for global health. *Healthcare (Basel)* 11(2), 242.

Kandasamy, J., Kinare, Y.P., Pawar, M.T., Majumdar, A., Vimal, K.E.K. *et al.* (2022) Circular economy adoption challenges in medical waste management for sustainable development: An empirical study. *Sustainable Development* 30(5), 958–975.

Kane, S., Gandidzanwa, C., Mutasa, R., Moyo, I., Sismayi, C. *et al.* (2019) Coming full circle: How health worker motivation and performance in results-based financing arrangements hinges on strong and adaptive health systems. *International Journal of Health Policy and Management* 8(2), 101–111.

Khan, B.A., Cheng, L., Khan, A.A. and Ahmed, H. (2019) Healthcare waste management in Asian developing countries: A mini review. *Waste Management & Research* 37(9), 863–875.

Korkut, E.N. (2018) Estimations and analysis of medical waste amounts in the city of Istanbul and proposing a new approach for the estimation of future medical waste amounts. *Waste Management* 81, 168–176.

Lee, S.M. and Lee, D. (2022) Effective medical waste management for sustainable green healthcare. *International Journal of Environmental Research and Public Health* 19(22), 14820.

Mazumder, M.A., Bhuiyan, M.A., Tunon, C., Baqui, A.H., Chowdhury, A.I. *et al*. (1997) An inventory of health and family planning facilities in Dhaka City. In: *MCH-FP Extension Project (Urban), Health and Population Division*. International Centre for Diarrhoeal Disease Research, Bangladesh.

Mubarak, R. (1998) Hospital environment in Dhaka. [Research paper] [Research paper]. Bangladesh: BCAS.

Pereno, A. and Eriksson, D. (2020) A multi-stakeholder perspective on sustainable healthcare: From 2030 onwards. *Futures* 122, 102605.

Pruss, A., Giroult, E. and Rushbrook, P. (1999) *Safe Management of Waste from Healthcare Activities*. WHO, Geneva, Switzerland.

Rahman, A. and Saidur, K.M. (1999) Situation assessment and analysis of hospital waste management (a pilot study). Director General of Health Services, Dhaka, Bangladesh.

Rahman, K.M. and Melville, L. (2000) An investigation into the conversion of non-hazardous medical wastes into biogas – A case study from the health and family planning sector in Bangladesh. *Processes* 11(5), 1494.

Rashbrook, P., Chandra, C. and Gayton, S. (2000) Starting healthcare waste management in medical institution: A practical approach [Working paper EUR/00/5021817]. WHO Regional Office for Europe, Copenhagen.

Rashbrook, P., Chandra, C. and Gayton, S. (2023) Starting healthcare waste management in medical institution: A practical approach. Copenhagen: WHO Regional Office for Europe.

Ravindra, K., Sareen, A., Dogra, S. and Mor, S. (2023) Appraisal of biomedical waste management practice in India and associated human health and environmental risk. *Journal of Environmental Biology* 44(4), 541–551.

Shammi, M., Rahman, M.M., Ali, M.L., Khan, A.S.M., Siddique, M.A.B. *et al*. (2022) Application of short and rapid strategic environmental assessment (SEA) for biomedical waste management in Bangladesh. *Case Studies in Chemical and Environmental Engineering* 5, 100177.

Sonone, S.S., Jadhav, S., Sankhla, M.S. and Kumar, R. (2020) Water contamination by heavy metals and their toxic effect on aquaculture and human health through food chain. *Letters in Applied NanoBioScience* 10(2), 2148–2166.

Sordi Schiavi, C., Soares, H. and Silva, T. (2021) Sustainable innovation and leadership in the treatment of medical waste in Porto Alegre/RS. *Revista de Administração Da UFSM* 14, 1010–1031.

Sujon, H., Biswas, T.K., Chowdhury, A. and Chowdhury, M.E. (2022) Medical waste management: An assessment of district-level public health facilities in Bangladesh. *Cureus* 14(5), 24830.

WHO (1998) Teacher's guide management of wastes from health care activities. Geneva, Switzerland.

WHO (2000) Hospital waste management problems and strategic solution, an insight. Geneva, Switzerland.

WHO (2018) Health care waste. Available at: https://www.who.int/news-room/fact-sheets/detail/health-c are-waste (accessed 17 July 2025).

Further Reading

Akther, N. (2000) Medical waste management review. In: Ahmed, M.F. (ed.) *Bangladesh Environment 2000: A Compilation of Papers of the International Conference on Bangladesh Environment-2000*. BAPA, Dhaka, Bangladesh.

Chartier, Y., Emmanuel, J., Pieper, U., Prüss, A., Rushbrook, P. *et al*. (1999) *Safe Management of Wastes from Health Care Activities*. WHO, Geneva.

Islam, M.M. (2023) The management of medical waste in Bangladesh: A policy and practices analysis. PhD dissertation, Teesside University, UK.

Rahman, M.H., Ahmed, S. and Ullah, M.S. (1999) A study on hospital waste management in Dhaka City. In: Pickford, J. (ed.) *Integrated Development for Water Supply and Sanitation, Proceedings of the 25th WEDC International Conference, Addis Ababa*, pp. 342–345 (Vol. August30–September 2).

Rahman, M.M., Bodrud-Doza, M., Griffiths, M.D. and Mamun, M.A. (2020) Biomedical waste amid COVID-19: Perspectives from Bangladesh. *Lancet Global Health* 8(10), e1262.

10 Troubled Waters: Assessing Causes, Consequences, and Innovations in Pollution Control

Vipan Rai and Amrik Singh*
Lovely Professional University, India

Abstract

Water pollution remains a critical global challenge, threatening ecosystems, public health, and economic activities. Industrial waste, agricultural runoff, and inadequate sewage management significantly contribute to surface, groundwater, and marine pollution. Bioremediation through genetically engineered microorganisms offers an eco-friendly solution to mitigate pollution, yet challenges remain in large-scale application. Effective regulatory frameworks and sustainable practices are crucial to addressing this crisis. This chapter explores water pollution sources, impacts, and innovative remediation techniques while discussing theoretical and practical implications. Future research should focus on optimizing biotechnological interventions for large-scale environmental cleanup.

10.1 Introduction

Water pollution is one of the most pressing environmental issues, significantly affecting aquatic ecosystems, human health, and economic activities. Rapid industrialization, urban expansion, and unsustainable agricultural practices have exacerbated water contamination worldwide. In developing nations like Sri Lanka and Bangladesh, poor waste management, excessive agrochemical use, and inadequate sewage treatment systems contribute to widespread pollution, leading to severe health consequences and environmental degradation. Surface water contamination, groundwater pollution, and marine ecosystem threats continue to escalate, demanding immediate intervention and sustainable solutions.

Industrial pollution plays a crucial role in water contamination. Heavy metal pollutants such as lead, mercury, cadmium, copper, and zinc enter water bodies through industrial effluents, bioaccumulate in aquatic organisms, and pose long-term ecological risks. Textile and dyeing industries, in particular, contribute high levels of chemical pollutants, affecting local communities and ecosystems. Similarly, agricultural runoff containing pesticides, fertilizers, and toxic organic compounds further deteriorates water quality. Inadequate enforcement of environmental regulations in many regions allows these harmful practices to persist, exacerbating the problem.

Bioremediation has emerged as a promising approach to mitigating water pollution, leveraging the capabilities of genetically engineered microorganisms to degrade pollutants

*Corresponding author: amrikmhm@gmail.com

effectively. Advances in genetic engineering enable the development of microbial strains capable of breaking down toxic compounds and heavy metals, offering a cost-effective and environmentally sustainable alternative to conventional chemical and physical remediation techniques. However, challenges such as microbial adaptability, gene stability, and large-scale implementation need further exploration. Additionally, integrating bioremediation with physical and chemical treatment methods can provide a holistic approach to water pollution management.

Regulatory frameworks for managing water pollution vary across countries. In India, strict policies under the Environment Protection Act regulate genetically modified organisms (GMOs) for bioremediation. The European Union enforces stringent risk assessments before approving GMO-based environmental applications, while the United States implements multiple regulatory mechanisms to control industrial and agricultural pollution. In contrast, many developing nations struggle with weak enforcement, inefficient waste management systems, and limited technological advancements.

Addressing water pollution requires a multi-dimensional strategy encompassing policy enforcement, technological innovation, industrial responsibility, and community awareness. Enhancing wastewater treatment infrastructure, promoting sustainable agricultural practices, and implementing stricter industrial regulations are essential steps. Furthermore, research in biotechnological advancements, such as genetically engineered bacteria for pollutant degradation, offers hope for sustainable water pollution management.

10.2   Literature Review

Harmful algal blooms (HABs) are increasingly recognized as a major environmental issue, with significant implications for aquatic ecosystems, public health, and economic activities. These blooms occur when algae – particularly toxin-producing species – grow excessively in water bodies, often leading to the depletion of oxygen, the release of harmful toxins, and the disruption of aquatic food chains. While HABs can occur naturally, anthropogenic activities such as nutrient pollution and climate change have intensified their frequency and severity (Bhalla *et al.*, 2023; Singh and Bathla, 2023; Ansari and Singh, 2024; Singh and Ansari, 2024; Singh *et al.*, 2024; Singh and Singh, 2024). Rising global temperatures, shifts in precipitation patterns, and human-induced nutrient enrichment from agriculture and urban runoff have created favorable conditions for these blooms, making them a growing concern for environmental scientists and policymakers.

10.3   Climate Change as a Catalyst for HABs

Scientific research indicates that climate change is one of the primary drivers of HAB proliferation. Rising temperatures accelerate metabolic rates in algae, extending bloom durations and expanding their geographical range. Warmer waters also lead to increased stratification of lakes and oceans, reducing vertical mixing and creating stable, nutrient-rich surface layers that favor the growth of harmful algal species. In addition, altered precipitation patterns caused by climate change contribute to nutrient runoff from agricultural lands, increasing the levels of nitrogen and phosphorus in water bodies. These nutrients act as fertilizers for algal growth, leading to eutrophication and the dominance of harmful species such as *Microcystis*, *Karenia brevis*, and *Alexandrium* – known for producing potent neurotoxins (Bhalla *et al.*, 2023; Singh and Bathla, 2023; Ansari and Singh, 2024; Singh and Ansari, 2024; Singh *et al.*, 2024; Singh and Singh, 2024).

According to Griffith and Gobler (2020), climate change acts as a co-stressor in marine and freshwater ecosystems, intensifying the impact of harmful algal blooms. Warmer temperatures not only increase the growth rates of algae but also influence species composition, often favoring toxic and invasive species over benign ones. This shift in ecological balance further exacerbates the problem, making HABs more frequent and severe.

10.4   Ecological and Environmental Consequences

The ecological consequences of HABs are profound. They disrupt aquatic food webs by outcompeting beneficial phytoplankton and depleting oxygen levels through decomposition, resulting in hypoxic and anoxic conditions that lead to mass fish kills. Moreover, toxin-producing algae can cause bioaccumulation in marine organisms, impacting higher trophic levels, including humans. Coral reefs, seagrasses, and other critical marine habitats are also vulnerable to HAB-induced stress, further threatening biodiversity and ecosystem stability (Griffith and Gobler, 2020; Bhalla *et al.*, 2023; Singh and Bathla, 2023; Ansari and Singh, 2024; Singh and Ansari, 2024; Singh *et al.*, 2024; Singh and Singh, 2024).

10.5   Human Health Risks and Socio-Economic Impacts

HABs pose direct risks to human health through contaminated drinking water, recreational exposure, and seafood consumption. Toxins such as microcystins, brevetoxins, and domoic acid can cause severe illnesses ranging from gastrointestinal distress to neurological disorders. The economic impact of HABs is also substantial, affecting fisheries, tourism, and water treatment infrastructure. In the United States alone, economic losses attributed to HABs exceed $82 million annually due to declines in fisheries, medical expenses, and increased water treatment costs (Griffith and Gobler, 2020; Bhalla *et al.*, 2023; Singh and Bathla, 2023; Ansari and Singh, 2024; Singh and Ansari, 2024; Singh *et al.*, 2024; Singh and Singh, 2024).

10.6   Mitigation Strategies and Future Research Directions

Addressing the HAB crisis requires a multifaceted approach combining scientific research, policy interventions, and sustainable environmental practices. Reducing nutrient pollution through improved agricultural management, wastewater treatment, and wetland restoration is critical in limiting bloom formation. Additionally, monitoring systems using satellite remote sensing and predictive modeling can help detect early bloom developments, allowing for timely

mitigation efforts. Future research should focus on understanding how climate-driven changes in ocean and freshwater dynamics will influence HAB trends and the development of eco-friendly mitigation strategies such as algal bioremediation and biological control measures (Griffith and Gobler, 2020; Bhalla *et al.*, 2023; Singh and Bathla, 2023; Ansari and Singh, 2024; Singh and Ansari, 2024; Singh *et al.*, 2024; Singh and Singh, 2024). In conclusion, harmful algal blooms represent a growing global challenge exacerbated by climate change and human activities. Their far-reaching impacts on ecosystems, human health, and economies highlight the urgent need for proactive mitigation and policy reforms. By integrating scientific research with sustainable management practices, society can work toward minimizing the risks associated with HABs and ensuring the long-term health of aquatic ecosystems (Bhalla *et al.*, 2023; Singh and Bathla, 2023; Ansari and Singh, 2024; Singh and Ansari, 2024; Singh *et al.*, 2024; Singh and Singh, 2024).

Eutrophication has been widely studied due to its significant impact on aquatic ecosystems and water quality. The phenomenon is primarily driven by an excessive influx of nutrients, particularly nitrates and phosphates, from both natural and anthropogenic sources. Natural factors such as rainfall and flooding contribute to the accumulation of nutrients in water bodies, accelerating the process. Additionally, the aging of lakes leads to organic matter buildup, fostering cyanobacterial blooms and phytoplankton proliferation. However, human activities, including the intensive use of fertilizers and concentrated feeding operations in aquaculture, are the primary drivers of eutrophication (Sharma *et al.*, 2024). The runoff from agricultural lands, rich in nitrates and phosphates, leads to an overgrowth of aquatic vegetation, disrupting ecological balance. Similarly, wastewater discharge from industrial and animal feed operations further exacerbates nutrient loading, accelerating eutrophic conditions (Akinnawo, 2023; Bhalla *et al.*, 2023; Singh and Bathla, 2023; Ansari and Singh, 2024; Singh and Ansari, 2024; Singh *et al.*, 2024; Singh and Singh, 2024).

The consequences of eutrophication are profound and multifaceted. The most visible effect is the formation of algal blooms, which not only degrade water quality but also pose serious health risks. Algal overgrowth reduces light penetration, impeding the photosynthesis of submerged plants. The subsequent decomposition of algae leads to oxygen depletion, causing large-scale fish kills and the formation of hypoxic dead zones. The proliferation of cyanobacteria, often dominant in eutrophic conditions, results in the release of harmful toxins such as microcystins, which accumulate in aquatic organisms and pose risks to human and animal health. The ingestion of microcystin-contaminated water has been linked to liver damage, neurotoxicity, and an increased risk of colorectal cancer. Furthermore, the physical consequences of eutrophication, such as the clogging of irrigation systems, hindrance to transportation, and disruption of power generation, present additional environmental and economic challenges (Akinnawo, 2023; Bhalla et al., 2023; Singh and Bathla, 2023; Ansari and Singh, 2024; Singh and Ansari, 2024; Singh et al., 2024; Singh and Singh, 2024).

To mitigate the effects of eutrophication, various physical, chemical, and biological techniques have been explored. Physical techniques such as dredging and aeration have been widely used to remove nutrient-rich sediments and enhance oxygen levels in water bodies, thereby limiting internal nutrient cycling. Infiltration-percolation, a tertiary wastewater treatment method, has shown high efficiency in reducing nitrogen and phosphorus levels using layered filtration systems. Additionally, advanced membrane separation technologies, particularly nano-filtration combined with bioreactors, have demonstrated significant potential in reducing nutrient loads in wastewater. However, challenges such as membrane fouling persist, requiring further technological advancements (Akinnawo, 2023; Bhalla et al., 2023; Singh and Bathla, 2023; Ansari and Singh, 2024; Singh and Ansari, 2024; Singh et al., 2024; Singh and Singh, 2024). The ongoing research on eutrophication highlights the urgent need for sustainable mitigation strategies that integrate advanced treatment technologies with policy interventions. The implementation of nutrient management practices in agriculture, along with improved wastewater treatment, remains critical in preventing further degradation of aquatic ecosystems (Akinnawo, 2023).

Nanotechnology is becoming an important tool for treating water pollution. Traditional methods often fail to remove all pollutants, especially when multiple contaminants are present. Nanotechnology offers an efficient and low-energy solution to clean water for homes, agriculture, and industries. This is particularly useful in developing countries where water quality is a serious issue. Nanomaterials improve water treatment by removing harmful substances more effectively and sustainably (Naskar et al., 2022).

However, while nanotechnology has many benefits, it also poses risks to the environment. Some nanomaterials accidentally enter lakes, rivers, and oceans, where they can harm aquatic life. For example, carbon nanotubes (CNTs) settle in sediments and affect bottom-dwelling species. Studies show that single-walled carbon nanotubes (SWCNTs) can increase the death rate of aquatic larvae. Additionally, CNTs can disturb the balance of neurotransmitters like serotonin and dopamine in fish, which may affect their behavior and overall health (Naskar et al., 2022; Bhalla et al., 2023; Singh and Bathla, 2023; Ansari and Singh, 2024; Singh and Ansari, 2024; Singh et al., 2024; Singh and Singh, 2024). Graphene-based nanomaterials also raise concerns. Research shows that graphene oxide can damage cell membranes in green algae, produce harmful oxygen compounds, and cause cell death. Microalgae that consume graphene quantum dots experience stress and altered growth. In zebrafish embryos, exposure to high levels of graphene nanomaterials has been linked to developmental problems and reduced movement. While graphene is helpful for water purification, its uncontrolled release into water bodies can harm ecosystems (Naskar et al., 2022; Bhalla et al., 2023; Singh and Bathla, 2023; Ansari and Singh, 2024; Singh and Ansari, 2024; Singh et al., 2024; Singh and Singh, 2024). Silver nanoparticles (AgNPs) are widely used in water treatment because they kill harmful bacteria. However, when released into the environment, they can be toxic to aquatic organisms, especially those living at the bottom of water bodies. Research shows that AgNPs cause genetic damage, oxidative stress, and early death in some aquatic species. These nanoparticles also settle in sediments, where they accumulate and pose long-term risks to benthic life. The potential dangers of AgNPs highlight the need for strict monitoring and careful use of nanotechnology in water treatment (Naskar et al., 2022). Nanotechnology offers powerful solutions for cleaning polluted water, but its environmental impact must be carefully managed. While these advanced materials improve water quality, it is important to control their release into natural ecosystems to prevent unintended harm (Naskar et al., 2022).

10.7 Water Pollution and Management in South Asia

Climate change has severely impacted South Asia, increasing climate refugees and poverty. The region struggles with both political and technical barriers in addressing environmental challenges. Many decision-makers fail to acknowledge the seriousness of water pollution, leading to delays in sustainable solutions (Bhalla et al., 2023; Singh and Bathla, 2023; Ansari and Singh, 2024; Singh and Ansari, 2024; Singh et al., 2024; Singh and Singh, 2024). A unified approach across countries, focusing on prevention rather than damage control, is essential. Water resource management must balance conservation and development while ensuring that environmental costs are considered. However, bringing all stakeholders together remains a challenge, often leading to conflicts between development supporters and conservation groups. The ideal goal is to eliminate pollutant discharge into water bodies, but complete achievement is not always cost-effective (Sarker et al., 2021; Bhalla et al., 2023; Singh and Bathla, 2023; Ansari and Singh, 2024; Singh and Ansari, 2024; Singh et al., 2024; Singh and Singh, 2024). A significant concern is the contamination of drinking water. Although most of South Asia's population has access to improved water sources, these are often polluted, leading to health risks. Unsafe water causes childhood deaths, reduced cognitive development, and economic struggles. Improving water quality requires coordinated efforts among water engineers, but there is often limited assessment of the health impact of interventions. Key challenges include groundwater depletion, surface water pollution, and inadequate sanitation.

Rural areas suffer the most, with open defecation and lack of toilet facilities creating severe public health risks. Countries like India, Pakistan, and Nepal experience significant urban-rural disparities in sanitation, highlighting deep social inequalities (Sarker et al., 2021; Bhalla et al., 2023; Singh and Bathla, 2023; Ansari and Singh, 2024; Singh and Ansari, 2024; Singh et al., 2024; Singh and Singh, 2024).

Water should be treated as both a public resource and an economic good. While economic principles can guide efficient water distribution, essential domestic water use should not be subject to market-driven pricing. Developing countries rely on subsidies for water infrastructure, making full cost recovery difficult. Ensuring water access for the poor requires a balance between economic efficiency and social responsibility. River basin management is another critical aspect of water governance. A successful strategy must integrate hydrological, political, and economic factors. However, in South Asia, water management agencies often work in isolation, leading to inefficiencies. A comprehensive river basin approach should include trained personnel, proper data collection, and active community participation (Sarker et al., 2021; Bhalla et al., 2023; Singh and Bathla, 2023; Ansari and Singh, 2024; Singh and Ansari, 2024; Singh et al., 2024; Singh and Singh, 2024). Recycling wastewater and industrial effluents can reduce pollution while conserving freshwater. Some Indian cities already use treated sewage for irrigation and sanitation, but large-scale implementation is still lacking. Community involvement in water and sanitation projects can be effective, particularly in underserved urban areas. Governments should encourage participation by creating formal mechanisms for engagement. Long-term solutions to water pollution in South Asia require coordinated regional efforts, investment in waste management, and prioritization of sustainable water policies (Sarker et al., 2021).

Water is an essential resource for both natural ecosystems and human activities. However, factors such as population growth, industrialization, urbanization, and climate change have contributed to the increasing contamination of water sources. Access to clean water remains a significant challenge, particularly in developing countries. Water pollution primarily originates from domestic waste, agricultural runoff, industrial discharges, excessive pesticide and fertilizer use, and rapid urban expansion. Contaminated water poses serious health risks as it facilitates the spread of bacterial, viral, and parasitic diseases. Therefore, ensuring water purification before consumption is crucial (Bhalla *et al.*, 2023; Singh and Bathla, 2023; Ansari and Singh, 2024; Singh and Ansari, 2024; Singh *et al.*, 2024; Singh and Singh, 2024). Natural coagulants have gained considerable attention due to their biodegradability and environmental safety. Among them, *Moringa oleifera* stands out for its effectiveness in water treatment. Native to India but now cultivated globally, *Moringa oleifera* is a nontoxic, drought-tolerant plant with medicinal and nutritional benefits. Its seeds contain water-soluble cationic proteins that serve as natural coagulants, effectively reducing turbidity, chemical oxygen demand, and total dissolved solids in wastewater. *Moringa oleifera* achieves up to 99% turbidity removal and 98% heavy metal elimination from surface water. Additionally, it has antimicrobial properties, buffering capacity, and natural antioxidants that enhance its performance as a water treatment agent. Compared to synthetic coagulants, *Moringa oleifera* is cost-effective, noncorrosive, and biodegradable, making it a sustainable alternative to conventional treatment methods.

Traditional synthetic coagulants, including aluminum salts like alum, are widely used due to their efficiency and affordability. However, prolonged use of alum has been associated with neurotoxicity, environmental accumulation, and potential links to Alzheimer's disease. Furthermore, synthetic coagulants generate substantial sludge, increasing treatment and disposal costs. To address these issues, researchers advocate for eco-friendly alternatives such as plant-based coagulants (Bhalla *et al.*, 2023; Singh and Bathla, 2023; Ansari and Singh, 2024; Singh and Ansari, 2024; Singh *et al.*, 2024; Singh and Singh, 2024). Water coagulation involves aggregating suspended particles into larger flocs using coagulants, facilitating their removal through sedimentation. This process significantly reduces turbidity and dissolved impurities. Coagulants work by neutralizing the charge of colloidal particles, allowing them to clump together. The efficiency of coagulation depends on factors such as coagulant type, water properties, and optimal pH conditions. Water pollution arises from various sources, including industrial discharges, agricultural runoff, domestic waste, and natural processes like erosion. Contaminants such as heavy metals, pesticides, organic toxins, and pathogens pose severe risks to human health, contributing to diseases like cancer, respiratory disorders, and waterborne infections. Industrial pollutants, including cadmium, copper, and nickel, can cause kidney damage, liver disorders, and chronic respiratory issues. Ensuring effective water treatment is crucial to preventing these health hazards and safeguarding environmental sustainability (Mitiku, 2020; Bhalla *et al.*, 2023; Singh and Bathla, 2023; Ansari and Singh, 2024; Singh and Ansari, 2024; Singh *et al.*, 2024; Singh and Singh, 2024).

Genetic engineering methods provide significant opportunities for removing pollutants and toxins from the environment. Compared to conventional technologies, these methods are more cost-effective and environmentally friendly. However, various environmental factors influence the bioremediation of contaminated sites. Microorganisms require optimal conditions for maximum performance and have specific limits of adaptation to environmental factors. Several biochemical, microbiological, ecological, and genetic factors affect the rate of bioprocessing and biodegradation of contaminants by genetically engineered bacteria used in environmental cleanup. Researchers continuously discover new genes that can be used to create novel bacterial strains capable of breaking down pollutants and producing new synthetic compounds (Bhalla *et al.*, 2023; Singh and Bathla, 2023; Ansari and Singh, 2024; Singh and Ansari, 2024; Singh *et al.*, 2024; Singh and Singh, 2024). Despite their potential, recombinant bacteria face several challenges in pollutant treatment. In complex environments with multiple microbial interactions, only a small number of genetically modified bacteria contribute to toxin removal. One effective method for increasing biodegradation capability is using a plasmid with multiple operons instead of multi-plasmids, as compatibility issues often arise. The protoplast fusion technique has shown promising results in breeding biodegradation-engineered bacteria. However, recombinant bacterial strains produced through this method sometimes contain unnecessary or harmful genes, which can hinder the breakdown process. Ensuring proper regulatory measures for the safe containment and application of genetically modified organisms (GMOs) in bioremediation is essential.

Further research in bioremediation focuses on identifying and analyzing gene and protein sequences that efficiently eliminate contaminants. Although genomics, metabolomics, and proteomics contribute to potential solutions for specific pollutants, challenges remain in optimizing GMOs for large-scale environmental cleanup. Combining bioremediation with physical and chemical techniques presents a comprehensive strategy for removing pollutants, demonstrating the long-term potential of this approach (Bhalla *et al.*, 2023; Singh and Bathla, 2023; Ansari and Singh, 2024; Singh and Ansari, 2024; Singh *et al.*, 2024; Singh and Singh, 2024). The rapid advancements in genetic engineering have led to concerns about the safe conduct of research involving GMOs. In India, GMOs and their applications are regulated under the 'Rules for the Manufacture, Use, Import, Export and Storage of Hazardous Microorganisms, Genetically Engineered Organisms or Cells, 1989'. These regulations fall under the Environment (Protection) Act, 1986, and are enforced by the Ministry of Environment, Forest, and Climate Change, the Department of Biotechnology, and state governments. Regulatory oversight is conducted through multiple authorities, including the rDNA Advisory Committee, Institutional Biosafety Committee, Review Committee on Genetic Manipulation, and Genetic Engineering Appraisal Committee. Internationally, regulatory frameworks vary. The United States regulates GMOs under various laws, including the Federal Insecticide, Fungicide, and Rodenticide Act and the Toxic Substances Control Act. The European Union enforces strict scientific risk assessments before approving GMO-based products. The European Food Safety Authority has established guidelines for evaluating the safety of GMOs in food and feed products. Water pollution, particularly surface water contamination, poses a severe environmental challenge in Sri Lanka. Urban lakes and downstream river sections suffer from high pollution levels due to inadequate sewage collection, poor industrial effluent management,

and insufficient sanitary facilities in low-income settlements. These factors have far-reaching consequences, not only affecting local communities but also contributing to broader ecological degradation. While regulations such as the Environmental Impact Assessment (EIA) and the Environmental Protection Licenses (EPL) aim to control pollution, inadequate enforcement and monitoring allow harmful waste disposal practices to persist (Bhalla *et al.*, 2023; Singh and Bathla, 2023; Ansari and Singh, 2024; Singh and Ansari, 2024; Singh *et al.*, 2024; Singh and Singh, 2024). Groundwater resources in Sri Lanka face significant challenges due to overuse and contamination from agrochemicals in rural regions and sewage infiltration in urban areas. Given financial constraints, pollution prevention remains the most cost-effective strategy for maintaining groundwater quality. A well-structured management system with comprehensive monitoring mechanisms is essential to prevent groundwater depletion and contamination. Additionally, marine pollution threatens Sri Lanka's fisheries and tourism industries, necessitating immediate intervention to control and mitigate environmental damage. Recognizing that surface water, groundwater, and marine pollution are interconnected is crucial for developing an effective water management strategy. Since 1994, Sri Lanka has encouraged cleaner technologies in industrial processes. The National Industrial Pollution Management Policy, enacted in 1996, emphasizes pollution prevention at the source and promotes industrial clustering within designated estates. Regulations for hazardous waste management were introduced in 1996, alongside initiatives to restrict agrochemical imports and promote organic farming and integrated pest management. However, the lack of stringent regulations governing agrochemical usage, particularly pesticides, remains a concern (Bhalla *et al.*, 2023; Singh and Bathla, 2023; Ansari and Singh, 2024; Singh and Ansari, 2024; Singh *et al.*, 2024; Singh and Singh, 2024). Sri Lanka has nearly 50 legislative measures focused on water-related issues, with 20 government entities overseeing water resource management. However, a lack of coordination among these agencies hinders efficient water regulation. The National Water Supply and Drainage Board (NWSDB) monitors drinking water quality, while the Central Environmental Authority (CEA) enforces pollution control measures. In 1993, Sri Lanka implemented EIA regulations, classifying industries based on pollution levels and determining necessary permits.

The Kelani River is one of the most polluted water bodies in Sri Lanka, primarily due to industrial liquid waste discharge, agricultural runoff, and municipal waste. Approximately 3000 industries along the riverbanks require pollution control licenses, but water tests consistently show violations of critical safe water quality standards. Improper sewage disposal further exacerbates water pollution. While Colombo has a sewage system, untreated sewage is discharged into the ocean; and in rural areas, inadequate sanitation facilities lead to direct sewage disposal into water bodies. Addressing these challenges requires strict regulatory enforcement and sustainable industrial and agricultural practices. Industrial pollution poses a severe environmental threat to Bangladesh, significantly impacting public health, water quality, and air conditions. While the industrial sector is a vital contributor to the country's gross domestic product (GDP), its rapid expansion has led to increasing levels of pollution. Without effective control measures, Bangladesh faces the risk of an environmental catastrophe. Heavy metal contamination in river water has dire consequences for aquatic life and human health, as toxic elements such as mercury, lead, and cadmium accumulate in fish and subsequently enter the food chain. The bioaccumulative nature of these pollutants poses a long-term risk to the environment and public health.

Contaminated water is unsuitable for household use and presents a direct hazard to both aquatic ecosystems and human well-being. The presence of bacteria such as *Salmonella* and *Escherichia coli* in water sources exceeds safety standards, making the water unsafe for drinking or even washing without proper treatment. Exposure to these pathogens can result in gastrointestinal illnesses, including diarrhea, nausea, and fever. Additionally, high concentrations of toxic chemical pollutants, such as nitrogen, sulfur, and phosphorus compounds, contribute to eutrophication, a process that depletes oxygen levels in water bodies and disrupts aquatic ecosystems. Excessive nitrate exposure is linked to serious health risks, including respiratory

illnesses and methemoglobinemia, a condition that affects oxygen transport in the blood (Bhalla *et al.*, 2023; Singh and Bathla, 2023; Ansari and Singh, 2024; Singh and Ansari, 2024; Singh *et al.*, 2024; Singh and Singh, 2024). Industrial pollution in Bangladesh has also been closely associated with various health conditions. Research has identified skin diseases, diarrhea, and dysentery as prevalent illnesses linked to polluted water. Textile dyeing industries are particularly problematic, as the use of alkali-based chemicals such as soda ash and caustic soda contributes to the alkalinity of water bodies. Many local communities report skin irritations and allergic reactions upon direct contact with polluted water sources. Pollutants from industries enter rivers and wetlands through drainage channels, further aggravating environmental degradation. Air pollution, another major consequence of industrial expansion, significantly contributes to respiratory infections, heart disease, and lung cancer. Vulnerable populations, including children, the elderly, and individuals with pre-existing health conditions, are disproportionately affected. Long-term exposure to particulate matter is responsible for a substantial portion of pollution-related deaths in Bangladesh. Reports indicate that household and ambient air pollution contributed to over 173,500 deaths in 2019 alone. Among the leading causes of mortality attributed to air pollution are lung cancer, lower respiratory tract infections, chronic obstructive pulmonary diseases, ischemic heart disease, and stroke. Addressing these environmental challenges requires urgent regulatory intervention, sustainable industrial practices, and effective pollution control measures. Water pollution caused by transition metals such as copper and zinc has significant implications for ecosystems, biodiversity, and human health. Industrialization, agriculture, and poor waste management contribute substantially to the contamination of water bodies. Even in regions with strict regulations, such as Europe, areas of intense industrial and agricultural activity continue to show high metal concentrations. Developing regions experience even greater pollution due to weak environmental controls. Zinc exhibits high mobility in water due to its solubility in oxidizing conditions and its retention in the solid phase. Studies indicate that zinc has a lower sorption capacity compared to copper, primarily due to metallo-organic complex formations and variations in ionic strength. The environmental behavior of these metals depends on pH levels, as higher pH enhances the stability of zinc–organic matter complexes, reducing bioavailability. Copper and zinc are essential elements, yet excessive levels lead to toxicity in aquatic life. Organisms show varying uptake and accumulation rates of these metals, with fish serving as a reliable indicator of copper and zinc pollution due to their metabolic dependence on these elements. Elevated concentrations of these metals interfere with cellular functions, triggering oxidative stress, metabolic imbalances, and immune responses. While aquatic animals deploy defense mechanisms such as metallothionein regulation, mucus secretion, and lysosomal storage, prolonged exposure to high concentrations results in ecotoxicological risks (Bhalla *et al.*, 2023; Singh and Bathla, 2023; Ansari and Singh, 2024; Singh and Ansari, 2024; Singh *et al.*, 2024; Singh and Singh, 2024). Bioremediation strategies are crucial for addressing copper and zinc contamination. Natural processes like adsorption and precipitation help remove these metals from aquatic environments. However, these must be supplemented with advanced bioremediation techniques such as bioaccumulation and biosorption to meet regulatory standards. Future research should focus on innovative biotechnological solutions, including genetic modifications in plants and microbes to enhance their metal-accumulation capabilities. Additionally, emission reduction strategies should be developed to minimize the release of these metals into the environment.

Copper mobility in water is influenced by sediment composition, water chemistry, and environmental conditions. Clay minerals, metal oxides, and organic matter contribute to copper adsorption, while non-clay minerals like quartz can increase metal mobility. The adsorption of copper is typically described by the Langmuir isotherm model, whereas zinc follows the Freundlich isotherm model, indicating different adsorption behaviors. The spatial variability in adsorption capacity highlights the role of sediment characteristics in determining copper movement.

Copper complexation in water significantly affects its bioavailability and toxicity. Natural organic matter, particularly humic and fulvic acids, plays a dominant role in copper binding, reducing its free ion concentration and toxicity. In freshwater, Cu^{2+} is the predominant form, whereas marine environments favor Cu^+ due to different chemical conditions. Despite regulatory efforts, copper contamination remains a global concern, with varying concentrations influenced by local industrial and agricultural activities. Copper accumulation in aquatic organisms varies based on water type, species, seasonal fluctuations, and environmental conditions. In freshwater, fish primarily absorb metals through their gills, while in saltwater, absorption occurs through both gills and intestines. Phytoplankton, present in both environments, play a crucial role in transferring copper up the food chain. Seasonal variations significantly affect metal accumulation, as seen in studies where fish like tench, pike-perch, and carp in Lake Damsa exhibited fluctuating copper levels across different seasons. Similarly, research from Polish freshwater bodies, western Ukraine, and Lake Taihu in China revealed regional differences in copper bioaccumulation, often influenced by human activity, pollution control measures, and biological factors. In contrast, fish from highly polluted regions like Lake Geriyo in Nigeria and the River Nile in egypt exhibited significantly higher copper accumulation, which can adversely impact physiological functions such as growth, respiration, and immune responses. Certain species, such as mollusks, frogs, and turtles, can tolerate higher copper levels, making them valuable bioindicators for aquatic pollution. Bioremediation offers a promising approach to mitigating copper pollution, utilizing natural processes like adsorption, precipitation, and microbial action. Bacteria play a critical role through bioaccumulation, biosorption, and biomineralization. Some bacteria, like *Pseudomonas putida CZ1*, exhibit remarkable tolerance and efficiency in removing copper from water, aligning well with the Langmuir isotherm model. Other species, such as *Myrica esculenta* and *Michelia figo*, demonstrate high adsorption capacities for copper ions, indicating their potential for large-scale bioremediation. Additionally, microbially induced carbonate precipitation (MICP) has emerged as an alternative method for purifying metal-contaminated water. Despite advancements, site-specific testing remains essential, particularly in atypical waters with extreme pH levels. The ongoing development of biological remediation techniques underscores the importance of sustainable and effective strategies for managing copper pollution in aquatic ecosystems.

10.8 Research Methodology

Owing to the broad nature of this study, which aims to connect theory and practice, a post-positivist paradigm is considered appropriate. Currently, the study's goals are primarily achieved by using secondary data. Previous studies, research articles, industry reports, journals, magazines, and other pertinent publications are examples of these secondary sources.

10.9 Discussion

Water pollution remains one of the most pressing environmental challenges, with significant consequences for ecosystems, human health, and socioeconomic development. The hospitality and tourism industry, due to its reliance on natural water bodies for recreational and operational purposes, plays a dual role as both a contributor to and a victim of water pollution. The findings from recent studies highlight the complexities of water pollution, ranging from the role of harmful algal blooms (HABs) and eutrophication to advancements in water treatment technologies such as nanotechnology and natural coagulants. This discussion synthesizes these findings, emphasizing the interconnected nature of pollution sources, environmental consequences, and potential mitigation strategies (Bhalla et al., 2023; Singh and Bathla, 2023; Ansari and Singh, 2024; Singh and Ansari, 2024; Singh et al., 2024; Singh and Singh, 2024).

10.10 Water Pollution in the Hospitality and Tourism Industry

The hospitality sector contributes to water pollution through excessive water consumption, improper wastewater disposal, and the release of pollutants such as detergents, food waste, and microplastics. Resorts, hotels, and tourist activities near coastal and freshwater ecosystems exacerbate pollution levels by increasing nutrient loads, which can lead to eutrophication and HABs. Studies indicate that human-induced factors, particularly nutrient runoff from agriculture and urban wastewater discharge, have significantly increased the prevalence of algal blooms (Griffith and Gobler, 2020; Bhalla et al., 2023; Singh and Bathla, 2023; Ansari and Singh, 2024; Singh and Ansari, 2024; Singh et al., 2024; Singh and Singh, 2024). The tourism industry, particularly in coastal destinations, faces the direct consequences of HABs, such as beach closures, marine ecosystem degradation, and threats to public health. Eutrophication is another critical issue that affects water quality in tourism-centric destinations. The excessive influx of nutrients, primarily nitrogen and phosphorus, results in increased phytoplankton growth, which depletes oxygen levels and disrupts aquatic food chains (Akinnawo, 2023). The decline in water quality directly impacts tourism-dependent economies by reducing the appeal of destinations and increasing the cost of water treatment. Furthermore, toxins produced by cyanobacteria during algal blooms pose health risks for both tourists and local communities, affecting the seafood industry and public water supplies.

10.11 Climate Change and Its Role in Water Pollution

Climate change acts as a catalyst for worsening water pollution, especially through its influence on HAB proliferation and water scarcity. Rising global temperatures have been linked to increased algal bloom frequencies, extended bloom durations, and the dominance of toxin-producing species over benign algae (Griffith and Gobler, 2020). Additionally, shifting precipitation patterns intensify nutrient runoff from agricultural lands, further fueling eutrophication. This has been observed in regions where extreme weather events, such as heavy rainfall, lead to excessive nutrient loading in water bodies, accelerating pollution levels (Sarker et al., 2021; Bhalla et al., 2023; Singh and Bathla, 2023; Ansari and Singh, 2024; Singh and Ansari, 2024; Singh et al., 2024; Singh and Singh, 2024). For the tourism industry, climate change-driven water pollution presents a major threat to water-based recreational activities, including swimming, diving, and fishing. The degradation of coral reefs due to nutrient pollution and ocean acidification further affects marine tourism, leading to revenue losses and damage to biodiversity. Without effective mitigation strategies, the increasing frequency of water pollution events could significantly impact the long-term sustainability of tourism in vulnerable regions.

10.12   Technological and Natural Approaches to Water Purification

To combat the effects of water pollution, scientific advancements have introduced promising solutions, including nanotechnology and biological water treatment methods. Nanomaterials have gained attention due to their ability to efficiently remove pollutants from water, particularly heavy metals, organic contaminants, and pathogens. However, their unintended environmental consequences, such as bioaccumulation in aquatic organisms and potential toxicity to marine life, raise concerns (Naskar et al., 2022). Carbon nanotubes and graphene-based materials, despite their effectiveness in purifying water, have been shown to disrupt aquatic ecosystems by affecting the growth and survival of microalgae and fish species. These findings emphasize the need for strict monitoring and regulation of nanotechnology applications in water treatment to prevent unintended ecological damage. Alternatively, natural coagulants, such as Moringa oleifera, offer a sustainable alternative to conventional chemical treatments. Studies suggest that Moringa oleifera is highly effective in removing turbidity and heavy metals while being biodegradable and nontoxic (Mitiku, 2020). Compared to traditional coagulants like aluminum sulfate, which has been linked to potential neurotoxicity, plant-based alternatives provide a safer and cost-effective approach to water treatment. The adoption of such sustainable solutions within the hospitality industry could significantly improve wastewater management and reduce the sector's environmental footprint.

10.13   Policy and Management Strategies for Water Pollution Control

Effective water management requires a comprehensive policy approach that integrates scientific research, regulatory frameworks, and community participation. In South Asia, where rapid industrialization and urbanization have exacerbated water pollution, policy inefficiencies and lack of enforcement remain key barriers to sustainable water management (Sarker et al., 2021). The hospitality industry, as a major water consumer, has the potential to adopt best practices such as wastewater recycling, the use of eco-friendly cleaning products, and the implementation of green infrastructure for water conservation.

The concept of integrated river basin management, which considers hydrological, political, and economic factors, has been proposed as an effective strategy for addressing transboundary water pollution. However, fragmented governance structures and competing interests often hinder collaborative efforts. Successful water management models require active participation from all stakeholders, including local governments, businesses, environmental organizations, and indigenous communities (Bhalla et al., 2023; Singh and Bathla, 2023; Ansari and Singh, 2024; Singh and Ansari, 2024; Singh et al., 2024; Singh and Singh, 2024). In addition to policy reforms, corporate responsibility within the hospitality and tourism sector must be emphasized. Hotels and resorts should prioritize water efficiency measures, invest in advanced wastewater treatment technologies, and engage in eco-certification programs that promote sustainable water practices. Public awareness campaigns targeting tourists and local communities can further encourage responsible water use and pollution reduction efforts.

As water pollution continues to threaten aquatic ecosystems and economic activities, future research should focus on developing innovative and sustainable mitigation strategies.

More studies are needed to explore the long-term ecological effects of nanotechnology in water treatment and identify safer alternatives with minimal environmental impact. Additionally, research on climate-resilient water management strategies will be crucial in adapting to changing hydrological conditions and preventing further deterioration of water quality. Further investigation into bioremediation techniques, such as the use of microbially induced carbonate precipitation (MICP) and genetically engineered bacteria for pollutant degradation, could provide new insights into effective and eco-friendly water purification methods (Naskar et al., 2022). Similarly, integrating remote sensing technologies with predictive modeling could enhance early detection and management of harmful algal blooms, enabling timely intervention measures. The issue of water pollution is a growing global challenge. As tourism expands, especially in coastal and ecologically sensitive regions, improper waste disposal, sewage mismanagement, and industrial discharges exacerbate water contamination. In response, biotechnological innovations such as genetic engineering, microbial bioremediation, and advanced pollution control measures have emerged as potential solutions. However, despite advancements, challenges remain in implementing these techniques on a large scale. This discussion explores the effectiveness of genetic engineering in water pollution mitigation, the regulatory frameworks governing genetically modified organisms (GMOs), and the limitations of current pollution control measures in countries such as Sri Lanka, Bangladesh, and other industrialized and developing regions.

10.14   Effectiveness of Genetic Engineering in Bioremediation

Genetic engineering has revolutionized environmental cleanup through the creation of genetically modified bacteria capable of breaking down toxic pollutants. Recombinant bacterial strains enhance bioremediation by introducing specific genes responsible for pollutant degradation. However, their efficiency is often limited by environmental factors such as pH levels, temperature variations, and microbial competition. For

instance, while some engineered strains exhibit high bioremediation capabilities in controlled laboratory settings, their effectiveness in natural environments diminishes due to ecological constraints.

One of the most effective approaches in bioremediation involves using plasmid-based genetic modifications. By integrating multiple operons on a single plasmid, scientists have improved bacterial adaptation to toxic environments. The protoplast fusion technique, which enables the development of hybrid microbial strains, has also been successful in increasing biodegradation efficiency. However, challenges such as genetic instability, plasmid loss, and unintended metabolic changes in recombinant bacteria limit the widespread application of these techniques. Ensuring that genetically engineered bacteria function effectively outside laboratory conditions requires further research into gene stability, host compatibility, and long-term ecological effects.

Despite these advancements, ethical and biosafety concerns persist regarding the deliberate release of GMOs into ecosystems. Regulatory bodies have established stringent containment measures to prevent unintended ecological consequences. Nevertheless, scientific advancements must balance environmental sustainability with responsible genetic engineering practices to ensure long-term safety.

10.15   Regulatory Frameworks for GMOs in Water Pollution Control

The introduction of genetically modified microorganisms (GMMs) in bioremediation necessitates strict regulatory oversight to ensure ecological safety. In India, the governance of GMOs falls under the Environment (Protection) Act, 1986, with the Ministry of Environment, Forest, and Climate Change enforcing the 'Rules for the Manufacture, Use, Import, Export and Storage of Hazardous Microorganisms, Genetically Engineered Organisms or Cells, 1989'. The multitiered approval system, including the Genetic Engineering Appraisal Committee (GEAC), ensures that GMMs undergo rigorous risk assessments before environmental release.

Internationally, regulatory frameworks vary significantly. The United States enforces GMO regulations through multiple laws, including the Toxic Substances Control Act (TSCA) and the Federal Insecticide, Fungicide, and Rodenticide Act (FIFRA). Meanwhile, the European Union follows a more precautionary approach, mandating scientific risk assessments and long-term environmental monitoring before GMO approval. The European Food Safety Authority (EFSA) plays a crucial role in assessing GMO safety for food and environmental applications.

In contrast, developing nations such as Sri Lanka and Bangladesh struggle with enforcing stringent GMO regulations. While Sri Lanka has over 50 legislative measures addressing water-related issues, a lack of coordination among government entities hampers effective pollution management. Similarly, Bangladesh faces regulatory enforcement challenges, as industrial expansion often outpaces environmental governance. Without adequate policy integration and enforcement, the introduction of GMOs in bioremediation remains a contentious issue in these regions.

10.16 Challenges in Industrial and Agricultural Pollution Management

Water pollution from industrial and agricultural activities presents severe environmental and health risks. In Sri Lanka, rapid urbanization and inadequate sewage collection have led to the contamination of urban lakes and river systems. The Kelani River, one of the most polluted water bodies in Sri Lanka, suffers from extensive industrial liquid waste discharge, municipal waste dumping, and agricultural runoff. While pollution control licenses are mandated for industrial units along the river, ineffective regulatory enforcement leads to continued contamination (Bhalla et al., 2023; Singh and Bathla, 2023; Ansari and Singh, 2024; Singh and Ansari, 2024; Singh et al., 2024; Singh and Singh, 2024). Similarly, in Bangladesh, industrial pollution has caused significant water quality degradation. Heavy metal contamination from textile dyeing industries and manufacturing plants has introduced toxic elements such as

lead, mercury, and cadmium into river systems. These pollutants bioaccumulate in aquatic life, posing long-term health risks to humans and wildlife. In addition to heavy metal toxicity, microbial contamination from inadequate sewage treatment has led to outbreaks of water-borne diseases such as diarrhea and dysentery. Studies indicate that *Salmonella* and *Escherichia coli* levels in Bangladeshi water sources often exceed World Health Organization (WHO) safety limits, making the water unsuitable for consumption without extensive treatment (Bhalla et al., 2023; Singh and Bathla, 2023; Ansari and Singh, 2024; Singh and Ansari, 2024; Singh et al., 2024; Singh and Singh, 2024). Agricultural pollution, particularly from pesticide and fertilizer runoff, further exacerbates water quality issues. High concentrations of nitrogen, sulfur, and phosphorus compounds lead to eutrophication, which depletes oxygen levels in aquatic ecosystems and disrupts biodiversity. In Sri Lanka, the unregulated use of agrochemicals has led to concerns about groundwater contamination, particularly in rural areas. Despite regulatory initiatives to promote organic farming and restrict harmful pesticides, inadequate monitoring systems hinder effective enforcement.

10.17 Bioremediation Strategies for Heavy Metal Removal

Transition metals such as copper and zinc present significant challenges in water pollution management due to their bioaccumulative and toxic nature. In both industrialized and developing regions, high concentrations of these metals have been recorded in river sediments and aquatic life. The mobility and bioavailability of copper and zinc depend on factors such as water pH, sediment composition, and complexation with organic matter (Bhalla et al., 2023; Singh and Bathla, 2023; Ansari and Singh, 2024; Singh and Ansari, 2024; Singh et al., 2024; Singh and Singh, 2024). Natural bioremediation techniques, including adsorption, precipitation, and microbial action, play a vital role in mitigating heavy metal contamination. Certain bacterial strains, such as *Pseudomonas putida CZ1*, exhibit remarkable tolerance to copper and can effectively remove the metal through bioaccumulation and biosorption. Similarly, plant-based phytoremediation techniques using species such as *Myrica esculenta* have shown high adsorption capacities for metal ions, making them potential candidates for large-scale bioremediation projects.

In aquatic environments, seasonal variations influence metal accumulation in fish and other organisms. Studies conducted in freshwater lakes and polluted rivers indicate fluctuating copper and zinc levels based on human activity and pollution control measures. In highly contaminated regions, fish such as carp, tench, and pike-perch exhibit significantly higher copper accumulation, leading to physiological stress and metabolic disruptions. Utilizing mollusks and other aquatic bioindicators can aid in long-term monitoring of metal pollution trends.

10.18 Integrated Approaches for Sustainable Water Pollution Management

While genetic engineering presents promising solutions for environmental cleanup, integrating bioremediation with physical and chemical techniques enhances the overall effectiveness of pollution control strategies. Combining microbial remediation with advanced filtration systems, chemical precipitation, and adsorption technologies provides a more comprehensive approach to water quality restoration.

Developing nations must focus on strengthening regulatory enforcement, improving wastewater treatment infrastructure, and promoting sustainable industrial and agricultural practices. Encouraging public–private partnerships and investing in cleaner production technologies can help reduce pollution at the source. Additionally, expanding environmental education initiatives can increase awareness and community involvement in water conservation efforts.

In conclusion, genetic engineering offers significant potential for addressing water pollution challenges, particularly in bioremediation applications. However, ensuring the ecological safety of recombinant bacteria, strengthening regulatory frameworks, and integrating multidisciplinary pollution control strategies are crucial for achieving long-term environmental sustainability. Future research should prioritize optimizing genetically modified bacteria for real-world applications while simultaneously enhancing traditional water treatment methods to create a holistic approach to pollution mitigation.

10.19 Theoretical Implications

10.19.1 Bioremediation as an alternative to conventional water treatment

Bioremediation utilizes microbial processes to remove pollutants from water bodies, offering a sustainable and cost-effective alternative to traditional physical and chemical treatment methods. While conventional approaches such as coagulation, sedimentation, and chemical oxidation have limitations, genetically engineered microorganisms provide targeted and efficient contaminant degradation, reducing environmental harm.

10.19.2 Genetic engineering and microbial adaptability in polluted environments

Genetic modifications enhance microbial resistance to harsh environmental conditions, improving their ability to break down pollutants. However, the adaptability of engineered bacteria is influenced by microbial interactions, competition with native species, and environmental stress factors. Research in synthetic biology can optimize microbial efficiency for large-scale bioremediation.

10.19.3 Heavy metal bioaccumulation and ecosystem disruptions

Heavy metals such as copper, zinc, and lead bioaccumulate in aquatic organisms, causing toxicity at various trophic levels. The theoretical framework of bioaccumulation models helps in understanding how these pollutants persist in ecosystems and how bioremediation can be strategically implemented to minimize bioavailability and long-term environmental damage.

unintended environmental consequences, necessitating robust regulatory oversight.

10.19.4 Regulatory policies and global disparities in water pollution control

Different countries enforce varied regulatory mechanisms to manage water pollution. Theoretical analysis of international regulatory frameworks highlights the disparities in policy implementation, industrial compliance, and technological access, influencing the effectiveness of pollution control measures.

10.19.5 Microbial metabolism and genetic pathways for pollutant degradation

Understanding microbial metabolism is crucial for optimizing pollutant degradation. Genetic engineering allows researchers to manipulate metabolic pathways, enhancing the ability of microbes to degrade toxins. This theoretical approach aids in designing synthetic microbial strains for targeted pollution control.

10.19.6 Multiscale approach to pollution management

Water pollution research suggests integrating bioremediation with physical and chemical methods for comprehensive pollution management. A theoretical framework based on multiscale interventions can guide policymakers and industries in implementing efficient water treatment strategies.

10.19.7 Long-term ecological impact of engineered microorganisms

While bioremediation presents numerous advantages, concerns regarding the ecological impact of releasing genetically engineered microbes persist. Theoretical studies explore horizontal gene transfer, microbial mutation risks, and

10.20 Practical Implications

10.20.1 Enhancing industrial wastewater treatment

Industries should incorporate bioremediation techniques to treat wastewater before discharge. Utilizing genetically engineered bacteria capable of degrading industrial pollutants can significantly reduce environmental contamination and ensure compliance with regulatory standards.

10.20.2 Implementing sustainable agricultural practices

Agricultural pollution from pesticides and fertilizers can be minimized through bioremediation strategies such as bioaugmentation and phytoremediation. Encouraging organic farming and controlled chemical use can mitigate groundwater contamination.

10.20.3 Strengthening sewage and sanitation infrastructure

Poor sewage management contributes significantly to water pollution. Governments should invest in improved wastewater treatment plants, enforce sewage treatment mandates, and promote decentralized sanitation systems in rural areas.

10.20.4 Community awareness and environmental education

Public engagement is crucial in tackling water pollution. Conducting awareness programs on pollution prevention, responsible waste disposal, and water conservation can drive behavioral changes and foster sustainable practices.

10.20.5 Developing cost-effective bioremediation technologies

Financial constraints hinder the widespread adoption of bioremediation in developing nations. Research should focus on cost-effective microbial solutions that can be scaled up for large-scale water treatment applications.

10.20.6 Strengthening environmental regulations and compliance

Governments must enforce stricter environmental regulations, ensuring industries adhere to pollution control norms. Strengthening monitoring systems, imposing penalties for violations, and incentivizing sustainable practices are critical steps.

10.20.7 Integrating bioremediation with traditional water treatment methods

Combining bioremediation with conventional treatment methods, such as sedimentation and filtration, enhances pollutant removal efficiency. This integrated approach ensures effective water purification while maintaining ecological balance.

10.21 Future Research

Future research should focus on optimizing genetically engineered microbes for large-scale water pollution remediation. Advanced genetic techniques, such as CRISPR-based modifications, can enhance microbial efficiency in degrading complex pollutants. Additionally, interdisciplinary studies should explore the long-term ecological impacts of bioremediation, ensuring safe and sustainable implementation. Research should also investigate hybrid remediation approaches that combine biotechnological, physical, and chemical treatment methods for more effective pollution control. Policy-oriented studies can assess regulatory frameworks, guiding nations in adopting robust governance structures to manage water pollution. Finally,

collaboration between academia, industry, and government agencies will be essential in translating research into practical applications.

10.22 Conclusion

Water pollution is a pervasive and escalating environmental crisis that threatens ecosystems, human health, and economic stability worldwide. Rapid industrialization, intensive agricultural practices, and inadequate waste management have significantly contributed to the contamination of water bodies, necessitating urgent and sustainable interventions. The presence of heavy metals, organic pollutants, and microbial contaminants in surface and groundwater has far-reaching consequences, from disrupting aquatic biodiversity to increasing the prevalence of waterborne diseases. Addressing this crisis requires a multifaceted approach that combines technological innovations, stringent regulations, and active community participation. Bioremediation has emerged as a promising and environmentally sustainable solution to mitigate water pollution, leveraging genetically engineered microorganisms to break down pollutants effectively. This approach offers a viable alternative to conventional physical and chemical water treatment methods, which are often costly and environmentally intrusive. By integrating bioremediation with traditional methods, a more comprehensive and efficient strategy for pollution management can be achieved. However, challenges remain in large-scale implementation, including microbial stability, ecological safety, and public acceptance of genetically modified organisms (GMOs). Further research is essential to optimize microbial strains for enhanced pollutant degradation, ensuring minimal ecological risks and long-term sustainability.

Policy and regulatory frameworks play a crucial role in water pollution management. While developed nations have established stringent environmental regulations, many developing countries struggle with weak enforcement, lack of technological resources, and economic constraints. Strengthening regulatory oversight, promoting sustainable industrial and agricultural practices, and investing in modern wastewater treatment infrastructure are essential steps toward reducing water pollution. Governments must also incentivize industries to adopt environmentally friendly waste disposal techniques and support research in emerging pollution control technologies.

Beyond technological and policy interventions, raising public awareness is fundamental to addressing water pollution. Communities must be educated on the impacts of pollution and encouraged to adopt responsible water usage and waste disposal habits. Localized initiatives, such as rainwater harvesting, wetland conservation, and community-led cleanup drives, can contribute significantly to maintaining water quality. Moreover, fostering collaborations between academia, industry, and government agencies can bridge the gap between scientific research and real-world applications, ensuring

that innovative solutions reach the areas where they are most needed. Water pollution remains one of the greatest environmental challenges of the modern era, requiring immediate and sustained efforts at local, national, and global levels. Bioremediation presents a promising and scalable solution, but its success depends on continued research, regulatory support, and widespread implementation. Strengthening environmental policies, improving wastewater treatment infrastructure, and promoting responsible industrial and agricultural practices are essential to mitigating pollution. Additionally, community engagement and public education play a critical role in fostering long-term sustainability. Only through a collaborative and multidisciplinary approach can water pollution be effectively controlled, ensuring clean and safe water resources for future generations.

References

Akinnawo, S.O. (2023) Eutrophication: Causes, consequences, physical, chemical and biological techniques for mitigation strategies. *Environmental Challenges* 12, 100733.

Ansari, A.I. and Singh, A. (2024) Adopting sustainable and recycling practices in the hotel industry and its factors influencing guest satisfaction. In: Tyagi, P., Nadda, V., Kankaew, K. and Dube, K. (eds) *Examining Tourist Behaviors and Community Involvement in Destination Rejuvenation*. IGI Global. Hershey, PA, pp. 38–47. DOI: 10.4018/979-8-3693-6819-0.ch003.

Bhalla, A., Singh, P. and Singh, A. (2023) Technological advancement and mechanization of the hotel industry. In: Tailor, R. (ed.) *Application and Adoption of Robotic Process Automation for Smart Cities*. IGI Global, Hershey, PA, pp. 57–76. DOI: 10.4018/978-1-6684-7193-7.ch004.

Griffith, A.W. and Gobler, C.J. (2020) Harmful algal blooms: A climate change co-stressor in marine and freshwater ecosystems. *Harmful Algae* 91, 101590.

Mitiku, A.A. (2020) A review on water pollution: Causes, effects and treatment methods. *International Journal of Pharmaceutical Sciences Review and Research* 60(2), 94–101.

Naskar, J., Boatemaa, M.A., Rumjit, N.P., Thomas, G., George, P.J. et al. (2022) Recent advances of nanotechnology in mitigating emerging pollutants in water and wastewater: Status, challenges, and opportunities. *Water, Air, and Soil Pollution* 233(5), 156.

Sarker, B., Keya, K.N., Mahir, F.I., Nahiun, K.M., Shahida, S., and Khan, R.A. (2021) Surface and groundwater pollution: Causes and effects of urbanization and industrialization in South Asia. *Scientific Review* 7(3), 32–41.

Singh, A. and Bathla, G. (2023) Fostering creativity and innovation: tourism and hospitality perspective. In: Tyagi, P., Nadda, V., Bharti, V. and Kemer, E. (eds) *Embracing Business Sustainability through Innovation and Creativity in the Service Sector*. IGI Global, Hershey, PA, pp. 70–83.

Singh, A. and Ansari, A.I. (2024) Role of training and development in employee motivation: Tourism and hospitality sector. In: Mazurowski, T. (ed.) *Enhancing Employee Motivation Through Training and Development*. IGI Global, Hershey, PA, pp. 248–261. DOI: 10.4018/979-8-3693-1674-0.ch011.

Singh, V. and Singh, A. (2024) Digital health revolution: Enhancing well-being through technology. In: Nadda, V., Tyagi, P.K., Moniz Vieira, R. and Tyagi, P. (eds) *Implementing Sustainable Development Goals in the Service Sector*. IGI Global, Hershey, PA, pp. 213–219. DOI: 10.4018/979-8-3693-2065-5.ch016.

Singh, A., Tyagi, P.K. and Garg, A. (eds) (2024) *Sustainable Disposal Methods of Food Wastes in Hospitality Operations*. IGI Global, Hershey, PA. DOI: 10.4018/979-8-3693-2181-2.

Further Reading

Mahdi, A.J., Ewadh, H.M., Satee, S., Salman, J.M., and Rao, P.B. (2022) Comprehensive study of heavy metal pollution in surface water of Guntur region of Andra Pradesh. *IOP Conference Series: Earth and Environmental Science* 1088(1), 012016.

Nuwanka, M.W.R., and Gunathilaka, M.D.K.L. (2023) Complexities of water pollution: A review of surface water contamination in Sri Lanka. *International Journal of Environment, Engineering and Education* 5(2), 72–78.

Saha, A., Engelhardt, G.R., Macdonald, D.D., and Khan, R.A. (2023) Industrial pollution and its effect in the context of Bangladesh. *World Journal of Advanced Research and Reviews* 18(2), 936–945.

Sharma, S., Pathania, S., Bhagta, S., Kaushal, N., Bhardwaj, S., Bhatia, R.K., and Walia, A. (2024) Microbial remediation of polluted environment by using recombinant *E. coli*: A review. *Biotechnology for the Environment* 1(1), 8.

11 Industrial Waste and Environmental Impact: Trends, Challenges, and Sustainable Solutions

Ivneet Kaur Walia[1]*, Pyali Chatterjee[2] and Pranjali Gaur[1]

[1]*Rajiv Gandhi National University of Law, India;* [2]*Institute of Chartered Financial Analysts of India University (ICFAI), India*

Abstract

The study focuses on industrial waste, its meaning, its generation and its impact on the environment. In depth analysis of how in the last decade the waste generation has increased immensely, and its catastrophic impact on the pollution rates as well as the carbon footprint of all relevant nations and especially India. The chapter deals with minute intricacies related to methods employed by the various industries.

The study further delves into the upcoming and emerging technology that is being incorporated into waste management methods and weighs their pros and cons. Legal frameworks that are in place internationally as well as in India are also critically analysed highlighting the loopholes that continue to persist allowing corporations and industry moghuls to dump their waste illegally into non-suspecting countries, endangering human lives. In conclusion the research is summed up by providing criticisms as well as recommendations to put in place better equipment and measures for ensuring sustainable growth worldwide.

11.1 Introduction

The term 'industrial waste' is ascribed to the materials that have been brought into existence as the by-products of industrial processes and activities. These materials include various substances and residuary products that are intensively detrimental to the environment if they are disposed without proper procedure.

The 18th century marked the beginning of an industrial revolution in England, and soon after the world followed in its footsteps. For any economy to survive in the contemporary world, the government must prioritize industrialization. But the rapid industrialization and rampant use of resources to make those industries flourish, has left the world with a magnanimous amount of residue in all forms of materials, some toxic and others while nontoxic still hazardous for the environment.

It is the responsibility of each and every owner in that industry to ensure that the waste generated by their business is being disposed in the proper manner. Releasing industrial waste into public places or not being careful while disposing the waste, can lead to dangerous consequences for innocent people, whose resources as well as living space are contaminated. Hence, it is of utmost importance that industrial waste is disposed of with proper procedure, and the safety of the workers dealing with the waste is also ensured along with the general public.

*Corresponding author: ivneetwalia@rgnul.ac.in

This chapter aims to explore the intricacies and procedures that are in place to help with industrial waste disposal and explore the ways ahead in terms of technological advancement, community action, and government intervention on the waste management front.

The research also encompasses how various laws are currently in place to help with the waste disposal and holding the industries accountable for their messes. And concludes with recommendations.

11.2   Generation of Industrial Waste

Industrial waste is the collective waste of multiple and diverse industries and areas such as chemical factories, the food industry, the textile industry, mining operations, the pharmaceutical industry, and many others. These industries while performing their regular activities, produce a lot of excess material which needs to be disposed of quickly and in proper manner. This waste exists in different forms and types such as solid waste, liquid waste, wastewater, chemical waste, and electronic waste. Apart from these types, industrial waste can be categorized into different kinds such as hazardous and non-hazardous waste, biodegradable and non-biodegradable waste.

11.3   Types of Industrial Waste

11.3.1   Solid waste

The non-liquid excess by-products such as wood, paper, plastic, cardboard, food scraps, and other metals that are generated through the industrial processes and are left behind as residue are termed as solid industrial waste. This waste is either treated or is dumped in landfills and dump yards.

11.3.2   Liquid waste

The industrial liquid waste is the aquas wastes that are released into the environment by the industries. This waste consists of water in which

substances were dissolved and the water that was used during cleaning procedures, oils and greases from machineries and oils from the food industry, chemicals that were either produced as a result of failed experiments, and chemicals that were used in different stages of production.

11.3.3   Electronic waste

Waste consists of discarded electronic and electrical equipment such as computers, parts of television and other transmission parts as well as appliances, including mobile phones, Bluetooth devices, refrigerators etc. Electronic waste is increasing rampantly and is more difficult to dispose of than other wastes.

11.3.4   Hazardous and non-hazardous waste

Waste that has the capability to cause harm to humans, animals, and/or the environment is termed hazardous waste, it can be further categorized into toxic, corrosive, ignitable, and reactive. This type of waste cannot be simply disposed of in places where it can easily come in contact with living beings. it must be discarded at a facility established by the waste generators (captive facility) or at common hazardous waste treatment, storage, and disposal facilities (TSDFs) (Recykal, 2022).

Non-hazardous waste is the waste that does not show any potential to harm humans, animals, or the environment.

11.4   Waste Management Techniques

Every year India produces 62 million tons of municipal solid waste. This waste needs to be done away with the utmost carefulness and in such a manner that the environment and the people living in it are not exposed to life threatening situations due to this waste.

The industrial waste management techniques can be further divided in physical techniques, chemical techniques, and biological techniques. But before discuss the techniques, let us focus on a few simple techniques of industrial waste management.

11.4.1 Segregation

Just like the name suggests this strategy refers to separating recyclable waste from biodegradable, solid, hazardous, and non-hazardous waste. Recyclable waste such as glass, plastic, and paper could be segregated at the source and absorbed in the production or sold to authorized recyclers, but, sometimes, industrial wastes contain these recyclable components. This is where industrial waste management companies come in handy. These companies are specially equipped to deal with such industrial waste, freeing the industry of the time-consuming procedure of regulating waste.

11.4.2 Waste minimization

This is the practice of reducing the waste generated by the industries at the source level. In this strategy the industries actively undertake practices that reduce waste generation at the production level. This strategy can be implemented in three ways, namely:

1. Process optimization: in this phase the production processes are changed by putting in more efficient machinery and equipment, this helps in generating less waste at the source.
2. Materials used: in this phase the raw materials used by the industry are switched up to materials that be easily recycled or that can be reused for a different part of the production process or even in a different product that the industry manufactures.
3. Packaging material and strategy: in this phase the packaging material being used can be substituted by an easily biodegradable or reusable packaging so as to reduce any material waste and at the same time the manner in which the packaging is being done is also revised to curb any practice that may lead to the material being wasted.

11.4.3 Treatment

Treating industrial waste means stabilizing industrial waste to make it less hazardous before putting it in a landfill or composting makes it easier on the environment. Treatment involves changing the physical, chemical, or biological composition of a waste material through customized processes.

11.4.4 Recycling

Recycling means collecting, processing, and reusing materials that would otherwise be treated as waste. This involves the analysis of waste streams and production processes to find potential opportunities. By recycling (and beneficially repurposing) wastes to reduce disposal costs, and using or reusing recycled materials as substitutes, raw materials costs can be reduced.

11.5 Methods Used for Industrial Waste Management

11.5.1 Physical methods

I. *Immobilization/containment*: Containment is used where there is no need to remove the waste material and/or if the cost of removal is prohibitive. The main purpose of containment is to prevent or control liquid or semi-liquid contaminated wastes from leaking or leaching into surrounding non-contaminated areas. There are several basic techniques for such purposes including pumping, capping, draining, and the installation of slurry walls. The selection of various types of containment systems is based on the type of waste materials, and geohydrological conditions of the waste site. The role of site characteristics in containment system performance is underestimated. Often, more than one containment method is used in a given location. With time, the waste material may slowly biodegrade or chemically change to nontoxic forms or new treatment methods may become available for detoxifying the waste. In other words, containment may be used to 'buy time' under emergency conditions (Meegoda *et al.*, 2003; Tirumale *et al.*, 2016).

There are numerous types of containment systems. Their functional design approaches can be divided into two: the active and passive.

11.5.1.1 Active waste containment systems

II. *Disposal and pumping wells (including injection wells) and treatment technologies*: Flowable wastes can be pumped into porous or permeable subsurface rock that is protected with in surrounding layers of impermeable rock or soil (Meegoda *et al.*, 2003). Deep well injection is used extensively for the disposal of oil field brines. It may be used for by-products of industrial processes at locations near a suitable geologic formation. This method may only be applicable to non-hazardous waste such as oilfield brines, and it need regulatory approval. Since surrounding impermeable layers protect the disposal location, the only other caution with this method would be the contamination of upper layers, which may be avoided by having casing for the wells.

III. *Surface Impoundment and Storage Facilities*: A surface impoundment is a facility or part of a facility that is a natural topographic depression or human-made excavation that is designed to hold liquid wastes or wastes containing free liquids. Surface impoundments may be lined with earthen materials or polymeric layers. Examples of surface impoundments are holding, storage, settling, aeration pits, ponds, and lagoons. A lagoon is a shallow, usually artificial, pond used for temporary or permanent storage of liquids. A stabilization lagoon is a shallow pond for the storage of wastewater before discharge to a stream or to a treatment facility. A sump is a stationary device designed to contain an accumulation of hazardous waste resulting from a hazardous discharge from a tank, container, waste pile, surface impoundment, landfill, or other hazardous waste management structure (Meegoda *et al.*, 2003).

11.5.1.2 Passive waste containment systems

Typical examples of passive components include drain tile collection systems, barrier walls,

caps, and liners. Passive components require input of energy to maintain their operational mechanisms. Passive components can be further divided into three major subsystems.

IV. *Hydraulic barrier wall*: The control of contaminant migration from existing disposal sites or impoundments may necessitate vertical subsurface barriers. The walls can be made from different materials and can be put up in various ways and combinations.

Top seals surface seals, caps, covers are designed:
- to control surface water to maximize surface runoff and minimize infiltration into the landfill;
- to protect against burrowing animals that may dig into the landfill areas;
- to reduce odor and gaseous emissions; and
- to reduce leachate production and/or contaminant transport potential.

There are several types of materials that can be used for the top seal barrier layer, including natural clay, bentonite, synthetic membrane, membrane-soil composite, and soil-wastes (such as paper sludge and scrap rubber tire).

Bottom seals (liners): the barrier wall and top seal can be used in existing or a newly constructed containment system. However, the bottom seal is difficult to install underneath an existing waste or containment structure. In new waste containment facilities, it is necessary to provide a liner system beneath the waste disposal. The major purpose of a liner is to prevent leachate or waste from migrating into a non-polluted aquifer. The barrier layer for liner systems can consist of native clays, processed clays, and geosynthetic membranes previously discussed for top seals.

V. *Carbon absorption*: The major technical strategies highlighted for decreasing CO_2 emissions and fighting climate change include energy reduction, improving energy use efficiency, reducing the carbon intensity of energy sources, diversifying energy supply portfolio by promoting use of renewable, increasing biological CO_2 sequestration, and carbon capture and storage (CCS). CCS is a three step process consisting of (i) the separation of CO_2 from gaseous waste streams, (ii) the transport of CO_2 to storage locations, and (iii) the long-term isolation of CO_2 from the atmosphere (carbon storage or sequestration). CCS is attracting growing interest around the world, particularly in countries where electricity generation and export income are heavily dependent on fossil fuels, such as China, Western Europe, Australia, Canada, and the USA. CCS is viewed as an important bridging technology that will allow CO_2 emissions to be managed during fossil fuel dependence while society steadily increases the use of renewable energy. It has been surmised that CCS may contribute 15–55% of the cumulative global mitigation effort by 2100 and that the inclusion of CCS in a mitigation portfolio could reduce the costs of stabilizing atmospheric CO_2 concentrations by 30% or more (Bobicki, 2014).

VI. *Distillation*: Distillation, the ancient technique that separates elements of a liquid mixture by their boiling points turns out to be vital for both environmental and water treatment. Leveraging the mechanics of vaporization and condensation, distillation is a powerful and effective technique for water purification, pollutant removal, and resource recovery. The process involves heating a liquid solution, causing the most volatile components to vaporize. This vapor is then channeled through a condenser, where it cools and condenses back into a liquid. The collected condensate, now enriched with the desired components, is the distilled product, leaving behind the less volatile components in the original solution.

VII. *Filtration*: Industrial filtration is the process by which industries extract their products from the waste and by-products of production methods, along with other forms of waste. Filtration also emphasizes the polished final product and the efficient elimination and drying of waste for secure disposal. Industrial filtration systems come in various types and can function in a start stop manner or continuously based on the requirements of the business. There are different types of filters, such as:

* Gravitational filters: the most prevalent type of industrial filter, gravity filters, use natural gravitational forces to separate solids from liquids and eliminate debris and sediment from products. Through the application of strong gravitational pressure, devices such as centrifuges generating 3000 times gravitational force can differentiate materials based on density and efficiently remove waste from the final product.

* Filters for pressure and vacuum: rapidly emerging as the favored filtration technique across numerous sectors, pressure of vacuum filters utilize substantial force to pull materials through a fabric or refining medium, effectively separating them from other by-products.

* Leaf filtration systems: named for the slim metal plates contained within, leaf filters are the top option for edible oil and food sectors. Featuring vertical piles of as many as 60 leaves, apply pressure to strain the liquid through fine mesh and trap any particles included. The purified liquid is subsequently transferred to a storage tank as the by-product is eliminated.

* Rotatory vacuum filters with drums: employed in sectors that need continuous rotation for filtering high-viscosity slurries. Rotary drum vacuum filters serve as a solution for mining or renewable energy generation. While the drum spins, semi-vacuum pressure pushes the slurry through a fine mesh screen, forming deposits on the drum's surface. They are subsequently bundled for convenient, swift proposal.

11.5.2 Chemical methods

11.5.2.1 Stabilization

In recent years, techniques for solidification and stabilization have surfaced as effective alternatives for waste management. At present, many credible research institutions and professional firms are involved in waste stabilization and solidification methods. Regrettably, stabilization and solidification have been utilized as synonyms. The truth is that these are two distinct processes. Stabilization is a chemical method employed to make waste less toxic or less dangerous to the nearby environment. It lowers the risk associated with the waste itself. Instances of stabilization methods encompass the ion exchange of heavy metals within the aluminosilicate structure of a cement-based stabilization agent and the adsorption of heavy metals on to fly ash in aqueous environments. Solidification is utilized to convert the waste (as is) into a lasting and appropriate physical state that is more suitable for storage, landfill, or reuse. This can be done with or without chemical fixation. It establishes obstacles between waste elements and the environment by either lowering the waste's permeability, diminishing the effective surface area available for diffusion, or both (Veeresh *et al.*, 2010).

Chemical stabilization processes are used to treat industrial and hazardous wastes. These processes represent alternatives to mine disposal, ocean disposal, or conventional land filling. Most of these processes originated in the field of radioactive waste control and management. Chemical solidification and stabilization refer to waste treatment that results in the combined effect of:

1. improvement of physical properties, mechanical (stabilization);
2. encapsulation of pollutants (immobilization by fixation); and
3. reduction of solubility and mobility of the toxic substances (immobilization by isolation).

Mechanical stabilization reduces the bulk mobility of the waste and makes it durable and dimensionally stable. The final product should be capable of resisting deterioration caused by mechanical or environmental stresses. The extent of mechanical stabilization is evaluated by conducting strength and durability tests. Immobilization by fixation minimizes the local mobility of individual contaminant components. Leaching tests are conducted to determine the effectiveness of fixation. Immobilization

by isolation limits the risk of the contaminants reaching the boundaries of the deposited mass by providing protection against internal contaminants transport (avoiding internal and external cracking) and by restricting internal transport pathways (low permeability).

There are many solidification and stabilization techniques used in the industry for different types of waste materials. Each of these techniques is usually recommended for use with specific waste constituents. Hence, process selection involves weighing advantages and disadvantages for each particular project. Generally, stabilization/solidification techniques can be divided into the following six major categories:

1. Cement-based techniques
2. Pozzolanic techniques (silicate-based techniques)
3. Thermoplastic techniques (including bitumen, paraffin, and polyethylene incorporation)
4. Sorbent techniques
5. Organic polymer techniques
6. Encapsulation techniques (jacketing) (Wise and Trantolo, 1994)

11.5.2.2   Incineration

Incineration, a method for thermal waste treatment, entails regulated burning to diminish waste volume and possibly recover energy, transforming organic waste into ashes, heat, and combustion gases.

The principal gaseous products of waste incineration, like other combustion processes, are carbon dioxide and water vapor. And, like many combustion processes, incineration also produces by-products such as soot particles and other contaminants released in exhaust gases, and leaves a residue (bottom ash) of incombustible and partially combusted waste that must be emptied from incinerator chambers and properly disposed. The composition of the gas and ash by-products is determined, at least in part, by the composition of the wastes fed into an incineration facility. This feed stream composition can be altered by other waste management activities, such as reducing the amount of waste generated, reusing materials, and recycling waste materials for use as feedstocks for various manufacturing processes.

The exhaust gases from waste incineration facilities may contain many potentially harmful substances, including particulate matter, oxides of nitrogen, oxides of sulfur, carbon monoxide, dioxins and furans, metals such as lead and mercury, acid gases, volatile chlorinated organic compounds, and polycyclic aromatic compounds. Optimal conditions in a combustion chamber must be maintained so that the gases rising from the chamber mix thoroughly and continuously with injected air; maintaining the optimal temperature range involves burning of fuel in an auxiliary burner during start up, shut-down, and process upsets. The combustion chamber is designed to provide adequate turbulence and residence time of the combustion gases.

Operation of the incinerator also affects the emission of heavy metals, chlorine, sulfur, and nitrogen that may be present in the waste fed into the incinerator. Such chemicals are not destroyed during combustion. Mercury and its compounds, for example, are volatile, so most of the mercury in the waste feed is vaporized in the combustion chamber (National Research Council *et al.*, 2000).

11.5.3   Biological methods

11.5.3.1   Aerobic and anaerobic treatment

Aerobic wastewater treatment is a biological method that employs oxygen to decompose organic contaminants along with pollutants such as nitrogen and phosphorus (Show and Lee, 2017). Oxygen is constantly infused into the wastewater or sewage by a mechanical aeration system, like an air blower or compressor. Aerobic microbes consume the organic matter in the wastewater, transforming it into carbon dioxide and biomass that can be eliminated.

Various technologies exist for the aerobic processing of wastewater and sewage (Shah, 2024). These consist of:

- Traditional activated sludge: aerobic micro-organisms decompose organic matter in an aeration tank. This creates biological flocs (sludge) that are later removed from the treated water in a sedimentation tank.
- Moving bed biofilm reactor (MBBR): biofilm develops on plastic media that

are suspended and circulated within an aeration tank. These are held in the tank by retention screens.
- Membrane bioreactor (MBR): a sophisticated method that merges the activated sludge process with membrane filtration.

Aerobic wastewater treatment is a reliable, straightforward, and effective method that results in superior secondary effluent. The resulting sludge is scentless and can be marketed as superior agricultural fertilizer. When used alongside anaerobic treatment, aerobic treatment systems guarantee the total elimination of contaminants and nutrients (Veolia, 2019). This signifies that your wastewater can be released safely without violating strict environmental regulations.

Anaerobic treatment is a process to generate energy, in contrast to aerobic systems that generally use up more energy for their aeration purposes (Kassab *et al.*, 2010). It is not very technologically complex and is moderately priced technology which occupies less space and produces less excess sludge than the conventional aerobic treatment technologies. Net energy production from biogas makes the anaerobic treatment technology an attractive option over other treatment methods. Anaerobic digestion seems to be the optimum option for the treatment of high strength organic effluents (Ersahin *et al.*, 2011). Anaerobic technology has improved significantly in the last few decades especially in the treatment of industrial wastewaters. High organic loading rates can be achieved at smaller footprints by using high-rate anaerobic reactors for the treatment of industrial effluents (Chan *et al.*, 2009; Ersahin *et al.*, 2011).

11.6   Legal Framework Governing Industrial Waste Management in India

11.6.1   The environment (Protection) Act of 1986

The Environment (Protection) Act of 1986 is an extensive law that was created with the intention of safeguarding and enriching India's environment (Prasad, 2006).

Section 3(1): Gives the government of India the power to take action to safeguard and protect the environment, including the handling and discarding of the industrial waste.

Section 5: allows the federal government to commission any officer or person to take the steps that are needed to curb pollution and industrial waste.

Section 6: the government is empowered under this section to impose any restrictions necessary to control waste and waste management.

11.6.2   The hazardous waste (Management, Handling, and Transboundary Movement) Rules, 2016

It is one of the most important laws relating to the handling of hazardous industrial waste in India (Karthikeyan *et al.*, 2018). These regulations look over the protocols for handling, managing, and moving hazardous waste across India and internationally.

Rule 4 outlines the obligations of the facility's occupant with regard to the responsible management of hazardous waste. Labelling, packing, and keeping records are all part of this for appropriate waste management.

Rule 5 demands that businesses get permission from the State Pollution Control Board (SPCB) or Pollution Control Committee (PCC) before they can generate, handle, and dispose of hazardous waste.

Rule 6 lays out the protocols for securely storing and moving hazardous waste, outlining requirements for documentation, labelling, and packaging.

Rule 10 specifies how hazardous waste should be handled, disposed of, and recycled. It establishes specific treatment techniques (e.g. chemical treatment, incineration, and disposal in secure, approved landfills).

11.6.3   The manufacture, Storage, and Import of Hazardous Chemicals Rules, 1989

These rules cover the handling, importation, and storage of hazardous chemicals under the Environment (Protection) Act (Kumar *et al.*, 2021). By handling chemicals improperly, they hope to reduce the likelihood of industrial mishaps and environmental dangers.

Rule 4mandates that businesses dealing with hazardous chemical notify the government of the quantity and kind of chemicals used and/or stored by them.

Rule 7 dictates that proper safety measures to be followed by industries dealing with such hazardous chemicals.

11.6.4   The National Green Tribunal act, 2010

The National Green Tribunal (NGT) was set up to look after matters regarding industrial waste management, corporate social responsibility, and environmental degradation (Gill, 2014).

Section 14 empowers the NGT with jurisdiction in matters relating to hazardous waste management and protection of the environment.

Section 15 grants power to the tribunal to take action and punish those causing illegal and mismanaged handling of hazardous waste.

11.6.5   The public liability insurance act of 1991

The Public Liability Insurance Act of 1991 requires that businesses and industries have some sort of insurance coverage to ensure accountability for the waste generated and environmental degradation caused by them (Raghavan, 1997). It guarantees victims of workplace accidents compensation.

Section 6 calls for the setting up of an environmental relief fund.

11.6.6   Basel convention on the control of transboundary movements of hazardous wastes and their disposal

This convention governs the transportation of hazardous waste between nations and promotes disposal that is beneficial for the environment (Peiry, 2013).

Article 4 of this convention dictates that hazardous waste should not be exported to nations that do not have proper mechanisms of disposal.

It is stated in Article 9 of this convention that all the signatories must come together to prevent and curb illegal transportation of waste across the nations.

11.6.7 Stockholm convention on Persistent Organic Pollutants (POPs)

India is a signatory to the Stockholm Convention on Persistent Organic Pollutants (POPs), which aims to curb or do away with the use and production of POPs, POPs are hazardous substances that pile up on living things and are detrimental to the world at large (Mohapatra, 2021).

Article 3: nations are advised to limit the production and use of POPs.

Article 4: makes a case for ecologically sound ways to dispose of POP waste.

11.7 Recommendations

The major recommendations include:

- Prevention of waste be prioritized: it is high time that industries start minimizing the waste generated rather than finding ways to get rid of the waste that has been produced in extensive quantities. Waste minimizing is a crucial practice that must be adopted by every factory, plant and essentially be made a requisite for any industry to continue with their production.

- Another crucial step is to invest in developing infrastructure that would help to collect, sort, and then recycle the material. Industries should use a closed loop system in which the waste that is generated from the industry would be are reused in same process or different process.

- Opportunities for reusing industrial waste material should also be created that can be useful in other applications and processes.

- The industrial waste should also be treated properly and then disposed. Advanced treatment technologies should be adopted like advanced oxidation processes and biological treatment. The sites for waste disposal should be environment friendly and reduce health risks. Opportunities for exploring valuable resources from the waste streams should also be considered.

- There should be a shift from the linear economic model that is 'take-make-dispose' to a circular economic model where waste is viewed as a resource. Collaborations between industries, researchers, and policy makers in this regard should be encouraged, and also the benefits of adopting circular economy should be spread among people so that they have a better awareness about waste management.

References

Bobicki, E. (2014) Pre-treatment of ultramafic nickel ores for improved mineral carbon sequestration. PhD thesis. University of Alberta. Available at: https://era.library.ualberta.ca/items/472e5036-fa73-47f9-b383-aadc2aa19d76/view/5bbb7cec-f271-4e1a-9de3-5a01d7419cb7/Bobicki_Erin_R_201406_PhD.pdf (accessed 27 March 2025).

Chan, Y.J., Chong, M.F., Law, C.L. and Hassell, D.G. (2009) A review on anaerobic–Aerobic treatment of industrial and municipal wastewater. *Chemical Engineering Journal* 155(1–2), 1–18.

Ersahin, M.E., Ozgun, H., Dereli, R.K. and Ozturk, I. (2011) Anaerobic treatment of industrial effluents: An overview of applications. In: Einslag, F.S.G. (ed.) *Waste Water: Treatment And Reutilization.* IntechOpen, Rijeka, Croatia, pp. 9–28.

Gill, G.N. (2014) The national green tribunal of India: A sustainable future through the principles of international environmental law. *Environmental Law Review* 16(3), 183–202.

Karthikeyan, L., Suresh, V.M., Krishnan, V., Tudor, T. and Varshini, V. (2018) The management of hazardous solid waste in India: An overview. *Environments* 5(9), 103.

Kassab, G., Halalsheh, M., Klapwijk, A., Fayyad, M. and Lier, J.B. (2010) Sequential anaerobic–Aerobic treatment for domestic wastewater: A review. *Bioresource Technology* 101(10), 3299–3310.

Kumar, P., Maurya, P. and Henly, S. (2021) Regulatory framework on chemicals in India: Latest developments and challenges. *International Chemical Regulatory and Law Review* 4, 101–106.

Meegoda, J.N., Ezeldin, A.S., Fang, H.Y. and Inyang, H.I. (2003) Waste immobilization technologies. *Practice Periodical of Hazardous, Toxic, and Radioactive Waste Management* 7(1), 46–58.

Mohapatra, P. (2021) A critical review of effectiveness of the regulations in India on designated persistent organic pollutants (POPs) in the Stockholm convention. In: Kumari, K. (ed.) *Persistent Organic Pollutants*. CRC Press, Boca Raton, FL, pp. 181–200.

National Research Council, Commission on Life Sciences, Board on Environmental Studies and Toxicology, and Committee on Health Effects of Waste Incineration (2000) *Waste Incineration and Public Health*. National Academy Press, Washington, DC.

Peiry, K.K. (2013) The basel convention on the control of transboundary movements of hazardous wastes and their disposal: The Basel Convention at a glance. *Proceedings of the ASIL Annual Meeting* 107(2013), 434–436.

Prasad, P.M. (2006) Environment protection: Role of regulatory system in India. *Economic and Political Weekly* 41(13), 1278–1288.

Raghavan, V. (1997) Public liability insurance act: Breaking new ground for Indian environmental law. *Journal of the Indian Law Institute* 39(1), 96–115.

Recykal (2022) Recykal. *Hazardous and Non-Hazardous Waste*. Available at: https://recykal.com/blog/hazardous-and-non-hazardous-waste (accessed 18 July 2025).

Shah, M.P. (2024) *Aerobic and Anaerobic Microbial Treatment of Industrial Wastewater*. CRC Press, Boca Raton, FL.

Show, K.Y. and Lee, D.J. (2017) Anaerobic treatment versus aerobic treatment. In: Lee, D.-J., Jegatheesan, V., Ngo, H.H., Hallenbeck, P.C. and Pandey, A. (eds) *Current Developments in Biotechnology and Bioengineering*. Elsevier, Amsterdam, pp. 205–230.

Tirumale, K., Raina, R. and Vuppu, S. (2016) Immobilization technique of natural dyes, as a novel method to preserve industrially important *E. coli* and *Bacillus* species. *Journal of Applied Pharmaceutical Science* 6(5), 148–151.

Veeresh, M., Veeresh, A.V., Huddar, B.D. and Hosetti, B.B. (2010) Dynamics of industrial waste stabilization pond treatment process. *Environmental Monitoring and Assessment* 169(1), 55–65.

Veolia (2019) Aerobic wastewater treatment/Veolia water technologies UK. Veolia Water Technologies. Available at: https://www.veoliawatertechnologies.co.uk/technologies/aerobic-treatment (accessed 18 July 2025).

Wise, D.L. and Trantolo, D.J. (1994) *Process engineering for pollution control and waste minimization*. CRC Press, Boca Raton, FL.

12 Examining Challenges and Innovations in Enhancing Solid Waste Management in India's Mountainous Regions Amid Growing Tourism

Rohit Saroha[1]*, Prakash Chandra Pandey[2] and Sanjeev Kumar[3]

[1]*SGT University, India;* [2]*SRM University, India;* [3]*Lovely Professional University, India*

Abstract

Because of its challenging environment, complicated geography, and cold climate, managing solid waste is becoming increasingly difficult in distant locations, such as the India mountain region. This chapter investigates the state of solid waste management in the mountain region about growing tourism and potential environmental consequences in the area through personal conversations with key informants and published and unpublished literature. According to the study, a circular waste management system might be established by the united efforts of everyone connected to the mountain region at all levels. The findings highlight innovative approaches that can be adopted to improve waste management practices, including the implementation of a circular waste management system. In addition, the chapter details the scheduled and current operations for critical procedures such as waste source segregation, collection, utilization of material recovery facilities, and recycling that may contribute to the establishment of sustainable solid waste management in the mountain region and other comparable contexts.

Ultimately, this research underscores the necessity of collaborative efforts among local communities, government agencies, and tourism stakeholders to establish effective waste management strategies that protect the fragile mountain ecosystem while accommodating the demands of growing tourism.

12.1 Introduction

Managing the increasing amount of solid trash in mountainous areas of developing nations is a major concern (Pokhrel and Viraraghavan, 2005). In India, contamination of land and water bodies near human settlements is becoming a serious concern due to the strain of growing tourism and inadequate waste treatment systems. Tourism and waste management practices in hilly regions are linked to environmental degradation caused by open rubbish dumping and inadequate collection strategies. Furthermore, it is significantly more challenging to organize a waste management operation in a mountain community without disrupting the delicate ecosystem due to the rough topography of these areas (Byers *et al.*, 2019). In addition, poor solid waste management (SWM) causes soil quality degradation, drinking water contamination, and health issues (Shankar and Shikha, 2017).

The Himalayas are home to some of the tallest peaks on earth. The highest peaks in India are located in the Garhwal Himalaya,

*Corresponding author: saroharohit324@gmail.com

© CAB International 2025. *Municipal Waste Management: Policies and Strategies*
(eds S. Kumar *et al.*)
DOI: 10.1079/9781836990666.0012

the Karakoram ranges, and Kanchenjunga itself. These include Nanda Devi, Kamet, and Kanchenjunga (Gulia, 2007). Kanchenjunga is the tallest peak in India and the third-highest mountain in the world, rising to a height of 8586 meters (28,169 ft) above sea level. Every year, during the climbing season in April and May, numerous mountaineers strive to reach the summit of this mountain, which commands worldwide attention. Similar to this, hikers swarm to the Everest Base Camp in February through May for the pre-monsoon and in September through December for the post-monsoon seasons (Carter, 1983).

This has not only become the primary source of income for the local population, but it is also a key factor for development in the area. As tourism is the main livelihood strategy for the people of the Everest region, travel and trade-related business enterprises such as lodges, shops, restaurants, tea shops, and bakeries are established in almost every settlement. As of 2024, the travel and tourism industry in India accounted for 9.1% of the nation's gross domestic product (GDP), which exceeded $11.1 trillion (Hamid and Bano, 2023). Compared to the prior year, this is a 12.1% increase.

According to Wang (2017), in contrast to the mountain region's economic expansion, managing garbage is a concern since a significant amount of solid waste is produced year-round along trekking routes. Because of how widespread the issue is, the Everest Base Camp journey has been referred to as the 'world's highest junkyard' and called a 'garbage trail'. But owing to the efforts of all the stakeholders, the SWM situation in the area is improving. A collection point with multiple bins for different types of waste placed in a fenced and covered area was established at the Everest Next Interpretation Centre in Kanchenjunga, north-eastern India, with the intention of promoting source segregation and replacing landfills, given the growing number of landfills in the area.

Due to the existing inadequate infrastructure for waste disposal, the accumulation of trash and solid materials in the delicate alpine environment, directly and indirectly leads to ecological damage. Sustainable resource management is typically more vulnerable to increased human activity in the Himalayas'

inaccessible high places and protected ranges (Wani and Wani, 2019). This is especially true in India because, in contrast to other regions of the world where human settlement in national parks is discouraged, people and parks cohabit in India. Under such circumstances, ineffective SWM programs may cause ecological disturbances like soil pollution, surface water contamination, and other sanitary issues that result in illnesses or infections. They may also cause a decline in tourism and have a long-term negative economic impact on the region.

It has been ten years since the mountain region's garbage generation condition was extensively documented in scientific studies (Ding *et al.*, 2021). Paper, metals, and plastics are the region's three main issues, according to the study. A thorough investigation into the region's solid waste problems and potential solutions was published in many articles. It included details on the location of active landfills, standard SWM procedures, and pre-Covid solid waste management programs. This chapter seeks to evaluate the state of solid waste management (SWM) in the mountain region, present SWM practices, initiatives, and future plans, as well as the problems associated with implementing them. By doing so, it hopes to provide more ideas for potential waste management opportunities in the future.

12.2 Materials and Methods

Stok Kangri in Ladakh, is thought to be the highest trekking mountain in India, reaching up to 6126 meters (20,100 ft) (Orr *et al.*, 2017). Higher than the tallest mountains in Europe and Africa, it is also regarded as one of the world's most trek-able peaks.

The main population of the area, the Ladakh people, is equally well-known for their proficiency in mountaineering. India's wealthiest villages are found in the Galwan, Shyok, and Darbuk region, which contains towns like Stok Kangri, Mankarmo, Matho, and Hey.

12.3 Data Collection and Analysis

Using data on SWM in the Everest region both published and unpublished, this study was carried out through a survey of the literature. Initially, a complete investigation of the solid waste context in the area was conducted, considering the ecological disturbance caused by the reported solid waste activities. After reviewing previous solid waste management procedures and their methodology, more problems faced by current SWM operations were evaluated.

This chapter's discussion and findings are grounded in the literature review and in direct conversations with key players engaged in the SWM in the area.

12.4   Results and Discussion

The present Ladakh waste management system is predicated on moving waste from one location to another. The stakeholders involved in garbage management are working to create workable, affordable, and environmentally beneficial solutions. A major hazard to both human health and the environment appears to be posed by the typical approach to SWM, which includes open burning methods and poorly built waste pits that contaminate water sources and the air. Given that the trash pits' volume was less than the amount of waste produced, the waste handlers may have been forced to burn the material.

To better understand SWM in Ladakh, direct correspondence was carried out with the program development officer of the Housing and Urban Development Department (H&UDD) of the Union Territory of Ladakh (UT of Ladakh). The most active of these appears to be the Rural Development Department (RDD), a community-based organization founded and managed by the Stok Kangri local community (Srivastava *et al.*, 2015). Solid trash collection, separation, and treatment in the area is the responsibility of the RDD and other neighbourhood projects. RDD is thus constantly looking for the most effective means of achieving goals that will benefit both parties. RDD has been managing the environment surrounding the Everest region with due consideration of sanitation and hygiene, from monitoring illegal climbers to providing door-to-door garbage collection services in Mankarmo, Matho (with a minimum service charge). RDD has collaborated with various government and nongovernment bodies, private sectors, and

tourism stakeholders. Working in collaboration with Stok Kangri local community Next, a non-profit organization and based social enterprise, RDD is pursuing an innovative approach to solid waste management that involves the potential for up-cycling and recycling waste materials, thereby transforming trash into treasure through the creative application of waste materials.

By 2020, RDD had constructed two waste disposal facilities in Matho, five incinerators close to significant villages, nine organic/noncombustible waste collecting centers in major population areas, and five incinerators in the Everest region. Similarly, 34 restrooms had been built in multiple base camps and along the trekking track. Presently, RDD offers door-to-door rubbish collection services to 96 houses and several commercial businesses in Mankarmo, as well as 116 lodges, eateries, cafes, and shops in Stok Kangri; and the construction of an environmental station, a gathering place with several containers for various waste kinds (plastics, paper, glass, and organic garbage). For a considerable amount of time, waste gathered from Leh, the capital of Ladakh, and the surrounding area was transported to a landfill known as 'Bomb Guard'.

In Ladakh, those who opt out of the municipal door-to-door waste collection services can deposit their waste at designated environmental stations. The local authorities have established a pilot environmental station in Leh, serving as a model for future development across the region. Plans to expand these stations and phase out open dumping sites were delayed due to the COVID-19 pandemic. Additionally, a small material recovery facility (MRF) has been set up near Leh with a concrete base to prevent seepage and an enclosure to keep out wildlife. This facility has facilitated the segregation of valuable recyclables, including aluminum cans and glass bottles.

Burnable and non-burnable solid waste is the two categories into which the RDD divides the region's solid waste. Paper, plastic, textiles, clothing, and plastic bottles are examples of burnable garbage; however, non-burnable waste includes cans, tins, oxygen cylinders, glass bottles, batteries, medical supplies, and portable gas stoves with cylinders. Even though paper, plastics, and plastic bottles are classified as burnable waste by RDD, their classification may need to be reevaluated because burning them might release dangerous chemicals into the atmosphere. Depending on the characteristics and makeup of the material, different waste types may need varying degrees of treatment. To be recycled or used again, 42,000 kg of non-burnable garbage was moved from Stok Kangri to Khari in 2020 and 2022.

Waste management and sanitation are increasingly becoming a concern for the Leh municipality because of the city's fast urbanization, rising tourism, migrant labour inflow, and importation of different packaged foods. The Leh municipality has accomplished much in trash management and sanitation within the town, despite these growing difficulties. Leh erected a 30-ton solar-powered integrated waste management system in 2019 to address the problem (20 dry and 10 wet). There are two dust bins, one green and one blue, for the separate collection of dry and wet garbage given to each of the about 5700 households in Leh town. The municipality exclusively uses distinct vehicles, fitted with GPS for effective collection, to pick up separated rubbish from the source. A roster that takes into account the requirements of various communities, including night shifts, is followed when collecting. To hold awareness camps in each ward, the municipality works with the Ladakh Ecological Development Group, a nearby NGO. With a focus on the values of reduce, reuse, and recycle, as well as the significance of supplying the municipality with separated waste, these camps seek to educate the public about waste management. Furthermore, reusable carry bags have been provided to residents for free. Leh implemented a single-use plastic ban early on, outpacing other regions of the nation. Leh produces 8–10 tons of dry garbage and 2–4 tons of wet waste daily on average during the summer (May–September). Over the winter months of October through April, the average drops to 3–4 tons of dry waste and 1–1.5 tons of wet trash daily. The town finished 2017 with an open defecation free (ODF) designation and was ranked 70th out of about 500 towns in the north zone for cleanest town among those with less than 50,000 residents.

The SANGDA Project, which emphasizes cleanliness, is managed by the RDD in rural Ladakh. Ten locations with significant tourist traffic are served by this project, including solar colonies, Nimo, the area around Pangong Lake, Deskit, and Khatse.

12.4.1 Waste generation and characteristics in the Everest region

The sources of waste include local markets and businesses, medical and climbing equipment, culinary waste, and trekking and mountaineering supplies. Local laborers or mules are used to carry the solid trash produced by trekkers to disposal or collecting places specified by RDD. The non-burnable items are held separately in a pit close to the RDD station, and the burnable waste – such as paper and plastics – is burned once it has been transported to the collection location.

The number of domestic tourists visiting has increased from 55,685 in 2010 to 241,285 in 2019, exhibiting a compound annual growth rate of 17.69% (Dolma and Kumar, 2024). In contrast, the number of foreign tourists visiting has increased at a rate of just 6.40%.

The Leh Municipal Committee estimates that during the tourist season, 12 to 13 tons of rubbish is produced every day, consisting of 2–3 tons of wet waste and 9–10 tons of dry waste. An estimated 3–4 tons of dry rubbish are produced in Leh during the winter, when there are fewer people visiting the area.

During the first quarter of 2022, from January to March, the largest city in Ladakh, Leh, generated 244 tons of garbage, while Kargil, the second-largest city, generated 36 tons of rubbish. Leh produced 1033 tons of garbage and Kargil produced 759 tons in the second quarter. In Leh and Kargil, 658 tons and 593 tons of garbage, respectively, were produced during the third quarter; and 319 tons and 278 tons of garbage, respectively, were produced in Leh and Kargil during the fourth quarter.

Prior to the installation of the Leh waste management facility at Skampari in 2020, all of the city's waste was disposed of at a landfill known as 'Bomb Guard' outside of the city. This solar-powered facility has a daily capacity of 30 tons. Items such as plastic, tin, cardboard, paper, etc. are processed here along with dry garbage, and wet waste is composted.

Beyond leaving their imprints on the base camps and climbing pathways, tourists also add to the volume of solid waste generated. Because the main supply of drinking water in the area is melting snow at higher altitudes, the growing amount of human faeces in the area is causing health hazards to climbers and aggravating the waste management scenario. Since human waste is only visible when the snow melts, it is impossible to determine the precise amount of human waste because climbers dig holes in the snow to urinate. As of right now, the area is devoid of any rules regarding the management of human waste. Biodegradable bags with enzymes to break down human waste are used by certain climbers. The disintegration of human faeces in the bags takes around three to six months. The health of people downstream who depend on surface water for drinking is also at risk when human waste is disposed of in plastic bags and washed away by the river.

12.4.2 Implications of current and planned SWM activities

Improved SWM activities can significantly reduce land, water, and air pollution in Ladakh. Proper waste segregation, recycling, and disposal will minimize the contamination of natural resources, preserving the fragile ecosystem. The SWM practices help protect the region's unique biodiversity, including endangered species, by reducing the hazards posed by unmanaged waste. Ladakh, with its cold desert climate, is particularly vulnerable to climate change. By managing organic waste effectively (e.g. composting), greenhouse gas emissions can be reduced, contributing to climate resilience.

Better SWM reduces the incidence of diseases linked to poor waste management, such as respiratory issues and waterborne diseases. This leads to a healthier population. SWM initiatives often create jobs, from waste collection to recycling, benefiting the local economy. Skill development in waste management can empower local communities. Ladakh's unique cultural heritage can be protected by reducing the visual and environmental impact of waste

on cultural sites, maintaining the area's attractiveness to both locals and tourists (Singh and Kotru, 2018).

Ladakh's economy heavily relies on tourism (Vogel and Field, 2020). Effective waste management ensures cleaner surroundings, enhancing the region's appeal as a tourist destination and potentially boosting tourism revenue. Implementing recycling and composting programs can turn waste into resources, providing economic benefits through the sale of recycled materials and compost. Proper planning and execution of SWM can reduce long-term costs associated with waste management, such as the expenses linked to landfill maintenance and environmental remediation (Shekdar, 2009).

The need for effective SWM can drive improvements in local governance structures, ensuring better coordination among stakeholders, from municipal authorities to local communities. Engaging local communities in SWM planning and implementation fosters a sense of ownership, ensuring the sustainability of these initiatives. This can lead to stronger civic responsibility and community-led environmental stewardship.

A study revealed that placing trash cans at key locations along the trekking trail is an efficient approach to reduce garbage generation in hilly areas. The objective is to establish a well-managed integrated SWM system, which calls for both a social and scientific perspective. Encouraging residents of mountain villages to get waste management education would also assist in clarifying the purpose of integrated SWM and the significance of sustainable development (Senekane *et al.*, 2022). Programs that teach individuals how to correctly separate and dispose of their garbage, for instance, may help with material recovery.

Furthermore, the desire of nearby communities to decrease trash may result in the finding of commercial prospects, such as the conversion of garbage into valuable commodities. Reusing and recycling are preferred solutions in waste management since trash generation is connected to the life cycle of products and resources. However, depending on the technique employed and the required quality of the recovered items, recycling some products, like paper and glass, may require a lot of energy and water. Recyclable waste is mostly transported to Kathmandu for recycling because there are not enough resources or waste processing facilities in Ladakh. Most of the residual solid waste is burned.

Travelers who bring reusable water bottles that they can use for the duration of their visit could be one of the positive behavioral actions that could reduce the amount of solid waste produced. To effectively sort garbage at the source, waste management practitioners might apply additional behavioral nudge concepts. For example, sorting waste at the source is shown to be more successful when employing bins with name stickers on them and various colors. Research suggests that some architectural styles can improve the way that waste is collected and encourage the development of new habits. The trash problem might also be improved by implementing safe disposal procedures, composting, practical recycling techniques, and resource reuse whenever possible.

Taking everything into account, interested parties from various backgrounds should work together to find a workable waste management structure that is novel and appropriate for the Everest region. Visitors should be informed to bring sturdy metal bottles or other reusable containers to support their choice. Everyone visiting the area should see examples of environmentally responsible behaviour. Trekkers should continue carrying an empty plastic bottle until they come across a trash can, even though it may feel awkward at first. This encourages positive environmental behaviour. The burden of waste or empty bottles may be momentarily relieved by letting them fall to the ground, but people should be mindful of the long-term harm they may cause. Waste generally refers to 'unwanted', therefore adopting an ecologically conscious mindset can be as simple as realizing that we are accountable for the unwanted things and that they should be disposed of appropriately.

In mountainous places, there is an especially great need to raise awareness and support garbage education at all levels. Together, the mountain communities, locals, businesses, stakeholders, travel, trekking, and climbing enterprises, as well as small and big governments, should create a solid waste management strategy that is easily implementable with the resources at hand. By providing resources and aiding in the administration and enforcement of the waste plan, all sectors of society, including those who are directly or indirectly involved in the generation of waste in mountainous areas, should take responsibility for enhancing and modernizing SWM infrastructure. The implementation of small-scale solutions, like trash segregation at the source and alternative dumping alternatives that promote environmentally conscious behaviour, can effectively manage garbage in remote alpine locations.

12.5 Conclusion

Since tourism is the Everest region's main source of income and solid waste creation is increasing due to increased tourism, imposing a visitor quota may not be the best way to address the garbage problem. The Everest region's previous solid waste management techniques have increased the risk of disease, contaminated water bodies, and degraded the quality of the soil. This region is having trouble processing waste sustainably because of improper waste management. The RDD and its partners' waste management practices have helped to mitigate the waste problem to some level, but the solid waste issue still has to be addressed. Upgrading waste management is part of the RDD's future plan, which calls on all communities in the Everest region to get involved and voluntarily contribute to the solution. Although alternative measures, like raising awareness among hikers, backpackers, climbers, and residents, and treating garbage other than land filling and segregating it at the source, are being gradually implemented, they have not eliminated the habit of burning or incinerating waste frequently. If governments, corporations, and RDD collaborated, the Everest region might be able to attain sustained SWM using small-scale solutions that work well in isolated alpine environments. Since more tourists are anticipated, the Everest region needs to be ready with useful SWM strategies. Applying behavioral modification could improve garbage sorting procedures. In addition, trash to art design needs to be embraced to shape environmental behaviour. In the long run, repurposing rubbish will advance art and might even generate revenue for the neighbourhood. Additionally, one of the main steps toward achieving sustainable SWM in the mountain region should be the implementation of a circular system that includes source segregation, collection, and recycling.

References

Byers, A.C., Chhetri, N., Shrestha, M. and Gustafsson, T. (2019) Sustainable solid waste management in the Sagarmatha (Mt Everest) National Park Khumbu, Nepal,Khumbu, Nepal: Management plan. Washington, DC: Report prepared for the National Geographic Society.

Carter, H.A. (ed.) (1983) *The American Alpine Journal*, Vol. 25. The Mountaineers Books, New York.

Ding, Y., Zhao, J., Liu, J.-W., Zhou, J., Cheng, L. *et al.* (2021) A review of China's municipal solid waste (MSW) and comparison with international regions: Management and technologies in treatment and resource utilization. *Journal of Cleaner Production* 293, 126144.

Dolma, S. and Kumar, A. (2024) A bottom-up approach for sustainable cultural tourism in Ladakh: An initiative taken by women and homestays. In: Chica-Olmo, J., Vujičić, M., Castanho, R.A., Stankov, U. and Martinelli, E. (eds) *Sustainable Tourism, Culture and Heritage Promotion: Development, Management and Connectivity*. Springer Nature, Cham, Switzerland, pp. 39–51.

Gulia, K.S. (ed.) (2007) *Discovering Himalaya: Tourism Of Himalayan Region*, Vol. 2. ISHA Books, Delhi, India.

Hamid, S. and Bano, N. (2023) Coronavirus: Choking global and indian tourism economy and leaving industry on the ventilator. *Journal of Hospitality and Tourism Insights* 6(4), 1594–1617.

Orr, E.N., Owen, L.A., Murari, M.K., Saha, S. and Caffee, M.W. (2017) The timing and extent of Quaternary glaciation of Stok, northern Zanskar range, Transhimalaya, of northern India. *Geomorphology* 284, 142–155.

Pokhrel, D. and Viraraghavan, T. (2005) Municipal solid waste management in Nepal: Practices and challenges. *Waste Management* 25(5), 555–562.

Senekane, M.F., Makhene, A. and Oelofse, S. (2022) A critical analysis of indigenous systems and practices of solid waste management in rural communities: The case of Maseru in Lesotho. *International Journal of Environmental Research and Public Health* 19(18), 11654.

Shankar, S. and Shikha (2017) Management and remediation of problem soils, solid waste and soil pollution. In: *Principles and Applications of Environmental Biotechnology for a Sustainable Future*. Springer, Singapore, pp. 143–171.

Shekdar, A.V. (2009) Sustainable solid waste management: An integrated approach for Asian countries. *Waste Management* 29(4), 1438–1448.

Singh, V. and Kotru, R. (2018) *Report of Working Group II: Sustainable Tourism in the Indian Himalayan Region*. NITI Aayog, India.

Srivastava, V., Ismail, S.A., Singh, P. and Singh, R.P. (2015) Urban solid waste management in the developing world with emphasis on India: Challenges and opportunities. *Reviews in Environmental Science and Bio/Technology* 14, 317–337.

Vogel, B. and Field, J. (2020) (Re)constructing borders through the governance of tourism and trade in Ladakh, India. *Political Geography* 82, 102226.

Wang, B. (2017) *Waste and Sacredness: The Nature and Cultural Conception of Solid Waste in the Tibetan Areas in Southwest China*. University of Wisconsin-Madison, Ann Arbor, MI.

Wani, M. and Wani, S.M. (2019) Sustainability of Himalayan environment: Issues and policies. In: Peshin, R. and Dhawan, A.K. (eds) *Natural Resource Management: Ecological Perspectives*. Springer, Cham, Switzerland, pp. 31–45.

Further Reading

Kaseva, M.E. and Moirana, J.L. (2010) Problems of solid waste management on Mount Kilimanjaro: A challenge to tourism. *Waste Management & Research* 28(8), 695–704.

Manfredi, E.C., Flury, B., Viviano, G., Thakuri, S., Khanal, S.N. *et al.* (2010) Solid waste and water quality management models for Sagarmatha National Park and buffer zone, Nepal. *Mountain Research and Development* 30(2), 127–142.

Mir, A.A., Mushtaq, J., Dar, A.Q., and Patel, M. (2023) A quantitative investigation of methane gas and solid waste management in mountainous Srinagar City: A case study. *Journal of Material Cycles and Waste Management* 25(1), 535–549.

Mushtaq, J., Dar, A.Q., and Ahsan, N. (2020) Spatial–temporal variations and forecasting analysis of municipal solid waste in the mountainous city of north-western Himalayas. *SN Applied Sciences* 2, 1161.

Posch, E., Bell, R., Weidinger, J.T., and Glade, T. (2015) Geomorphic processes, rock quality and solid waste management – examples from the Mt Everest region of Nepal. *Journal of Water Resource and Protection* 7(16), 1921–1308.

Pradhan, U.M. (2009) Sustainable solid waste management in a mountain ecosystem: Darjeeling, West Bengal, India. PhD thesis, University of Manitoba

Rohit, M., Sethi, K., Khan, M., and Raina, A. (2023) Machine learning model for prediction of the chemicals harmfulness on staff and guests in the hospitality industry: A pilot study. *Data and Metadata* 2, 161.

Rohit, N.A., Bathla, G., Sethi, K., and Sharma, G. (2024) Harnessing artificial intelligence for a competitive edge in religious tourism: A case study of India. In: Nadda, V., Tyagi, P.K., Singh, A., and Singh, V. (eds) *AI Innovations in Service and Tourism Marketing*. IGI Global, Hershey, PA, pp. 397–410.

Saroha, R., Gusain, A., Verma, P., and Sain, K. (2024) Navigating challenges effective governance strategies for solid waste management in rural India: The Balli Qutubpur experience. In: Albattat, A., Singh, A., Tyagi, P.K., and Haghi, A.K. (eds) *Solid Waste Management for Rural Regions*. IGI Global, Hershey, PA, pp. 39–54.

Saroha, R., Bathla, G., Kumar, H., and Sharma, G. (2025) Towards waste minimization in food packaging: Exploring sustainable methods for packaging and disposal. In: Rana, V.S., Raina, A., and Bathla, G. (eds) *Global Sustainable Practices in Gastronomic Tourism*. IGI Global, Hershey, PA, pp. 191–204.

Sethi, K., Sharma, M., and Gusain, A. (2023) The acceptance of machine module in the hospitality industry with prospects and challenges: A review. In 2nd International Conference on Computational Modelling, Simulation and Optimization (ICCMSO), Bali, Indonesia, pp. 283–288.

Sharma, M., Bathla, G., Kaushik, A., and Rana, S. (2023) A study on impact of adaptation of AI-artificial intelligence services on business performance of hotels. In 2nd International Conference on Computational Modelling, Simulation and Optimization (ICCMSO), pp. 28–32.

Xiao, Y., Xiao, Q., Tan, H., and Luo, Y. (2020) Effects of mountain urbanization on greenhouse gas emissions from municipal solid waste management practices in southwest China. *Environmental Monitoring and Assessment* 192, 690.

Index